M336
Mathematics and Computing: a third-level course

GROUPS & GEOMETRY

UNIT GE2
PERIODIC AND TRANSITIVE TILINGS

Prepared for the course team by
Fred Holroyd

The Open University

This text forms part of an Open University third-level course.
The main printed materials for this course are as follows.

Block 1
Unit IB1   Tilings
Unit IB2   Groups: properties and examples
Unit IB3   Frieze patterns
Unit IB4   Groups: axioms and their consequences

Block 2
Unit GR1   Properties of the integers
Unit GR2   Abelian and cyclic groups
Unit GE1   Counting with groups
Unit GE2   Periodic and transitive tilings

Block 3
Unit GR3   Decomposition of Abelian groups
Unit GR4   Finite groups 1
Unit GE3   Two-dimensional lattices
Unit GE4   Wallpaper patterns

Block 4
Unit GR5   Sylow's theorems
Unit GR6   Finite groups 2
Unit GE5   Groups and solids in three dimensions
Unit GE6   Three-dimensional lattices and polyhedra

The course was produced by the following team:

Andrew Adamyk (BBC Producer)
David Asche (Author, Software and Video)
Jenny Chalmers (Publishing Editor)
Bob Coates (Author)
Sarah Crompton (Graphic Designer)
David Crowe (Author and Video)
Margaret Crowe (Course Manager)
Alison George (Graphic Artist)
Derek Goldrei (Groups Exercises and Assessment)
Fred Holroyd (Chair, Author, Video and Academic Editor)
Jack Koumi (BBC Producer)
Tim Lister (Geometry Exercises and Assessment)
Roger Lowry (Publishing Editor)
Bob Margolis (Author)
Roy Nelson (Author and Video)
Joe Rooney (Author and Video)
Peter Strain-Clark (Author and Video)
Pip Surgey (BBC Producer)

With valuable assistance from:

Maths Faculty Course Materials Production Unit
Christine Bestavachvili (Video Presenter)
Ian Brodie (Reader)
Andrew Brown (Reader)
Judith Daniels (Video Presenter)
Kathleen Gilmartin (Video Presenter)
Liz Scott (Reader)
Heidi Wilson (Reader)
Robin Wilson (Reader)

The external assessor was:
Norman Biggs (Professor of Mathematics, LSE)

The Open University, Walton Hall, Milton Keynes, MK7 6AA.

First published 1994. Reprinted 1997, 2002, 2005, 2009.

Copyright © 1994 The Open University

All rights reserved. No part of this publication may be reproduced, stored in a retrieval system or transmitted in any form or by any means, without written permission from the publisher or a licence from the Copyright Licensing Agency Limited. Details of such licences (for reprographic reproduction) may be obtained from the Copyright Licensing Agency Ltd of 90 Tottenham Court Road, London, W1P 9HE.

Edited, designed and typeset by the Open University using the Open University TEX System.

Printed in Malta by Gutenberg Press Limited.

ISBN 0 7492 2170 4

This text forms part of an Open University Third Level Course. If you would like a copy of *Studying with the Open University*, please write to the Central Enquiry Service, PO Box 200, The Open University, Walton Hall, Milton Keynes, MK7 6YZ. If you have not already enrolled on the Course and would like to buy this or other Open University material, please write to Open University Educational Enterprises Ltd, 12 Cofferidge Close, Stony Stratford, Milton Keynes, MK11 1BY, United Kingdom.

1.3

The paper used for this book is FSC-certified and totally chlorine-free. FSC (the Forest Stewardship Council) is an international network to promote responsible management of the world's forests.

CONTENTS

Study guide		4
Introduction		5
1	**Periodic tilings**	**5**
	1.1 Symmetry and translation groups of a tiling	5
	1.2 Translational tilings	9
	1.3 Translational orbits	12
2	**The Euler Equation**	**14**
	2.1 Translational Tiling Theorem	14
	2.2 The Euler Equation	16
3	**Orbit diagrams**	**18**
	3.1 Tile–edge diagram	18
	3.2 Vertex–edge diagram	19
	3.3 Tile–vertex diagram	20
	3.4 Using orbit diagrams	20
4	**Transitive tilings**	**23**
	4.1 Transitive group actions	23
	4.2 Transitivity and periodicity	24
	4.3 The Transitive Tiling Theorem	27
5	**The Grünbaum–Shephard classification**	**28**
	5.1 The incidence symbol of a transitive tiling	29
	5.2 Using incidence symbols	32
	5.3 The classification result	34
Appendix: proof of the Finitary Theorem		38
Solutions to the exercises		39
Objectives		49
Index		50

STUDY GUIDE

You should find that the study times for the sections are roughly similar, though the kind of work involved is quite varied. Sections 1, 3 and 5 contain numerous manipulative exercises; Sections 2 and 4, on the other hand, contain fewer exercises but four fairly challenging mathematical proofs. We have designed the unit in such a way that the more manipulative parts should help with the more abstract ideas in the proofs in Sections 2 and 4. Nevertheless, you should not be surprised or discouraged if you find Sections 2 and 4 quite demanding.

The video programme associated with this unit, VC2B, is designed to help you to gain familiarity with the concept of an *incidence symbol*, which is discussed and used in Section 5. The programme uses jigsaw puzzles in a way which should make the concept easier to visualize, and we advise you to view the programme after you have worked through Subsection 5.1 but before you start work on Subsection 5.2.

There is no audio programme associated with this unit.

You will be using many tiling cards and their overlays from the *Geometry Envelope* throughout Sections 1, 3 and 5 of this unit. Therefore, even if you have found it possible to study the previous units on a train or bus, it will not be feasible for this unit; you will definitely need to tidy your desk or table!

INTRODUCTION

In *Unit IB1* you met the concept of a *tiling*. We noted that, in general, a tiling need not have any recognizable pattern, but we concentrated on tilings that do have a pattern. It is after all impossible to represent a tiling of the whole plane adequately by drawing a finite portion of it, unless we assume that the structure in that portion does repeat itself in a predictable way.

This unit is concerned with tilings where there *is* an indefinitely repeating pattern. Now you have already seen (in *Unit IB3*) examples of another type of structure — a frieze — in which a basic motif repeats itself infinitely in one direction; but tilings cover the whole plane rather than just a strip, so here the requirement is that the repetition be in two independent directions. In this unit, we analyse such tilings in terms of their symmetry groups and the way these groups act on the parts of the tilings.

In Section 1 we formalize the idea of 'repeating pattern' by defining a *periodic* tiling and considering its translational symmetries. These symmetries form a group, and the tiles, vertices and edges fall into *orbits* under the action of this group. The section ends by noting a very simple equation — the *Euler Equation* — which establishes a relationship between the numbers of tile, vertex and edge orbits. The validity of this equation is demonstrated in Section 2.

Group actions were defined in *Unit IB2* and used extensively in *Unit GE1*.

In Section 3, we explore further the relationship between these orbits, by defining and using *orbit diagrams*. We then go on, in Sections 4 and 5, to consider *transitive tilings* (tilings where the *full* symmetry group — not merely the group of translational symmetries — produces just *one* tile orbit). The unit culminates in the classification of such tilings into 81 types. This is quite a recent result: the work on which Section 5 is based was first published in 1977.

1 PERIODIC TILINGS

1.1 Symmetry and translation groups of a tiling

In *Unit IB2*, we showed that the set of all symmetries of a geometric object constitutes a group. In particular, a tiling by rectangles such as that in Figure 1.1 has a symmetry group which contains translations in two independent directions (and is therefore an infinite group).

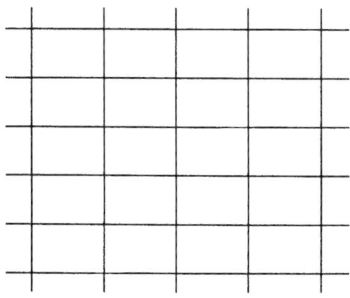

Figure 1.1

Translations are not the only symmetries of the above tiling; a rotation by π about any vertex, for example, is also a symmetry. However, the translations that are symmetries do form a group in their own right.

The situation is analogous to that which you saw for friezes in *Unit IB3*; the symmetry group $\Gamma(\mathcal{T})$ of a tiling $\mathcal{T}$ contains the group $\Gamma^+(\mathcal{T})$ of direct symmetries of $\mathcal{T}$, which in turn contains the group $\Delta(\mathcal{T})$ of translational symmetries:

$$\Gamma(\mathcal{T}) \supseteq \Gamma^+(\mathcal{T}) \supseteq \Delta(\mathcal{T}).$$

What form can $\Delta(\mathcal{T})$ take? Clearly, this group must be a subgroup of Δ, the group of *all* translations of the plane.

Now Δ has a wide variety of subgroups, but fortunately most of them need not concern us, since they contain arbitrarily small non-zero translations. Such a group clearly cannot be a candidate for the translation group of a tiling, as there must be a minimum size of any non-zero translation that maps whole tiles to whole tiles.

It turns out that there are essentially three possibilities for the translation group $\Delta(\mathcal{T})$ of a tiling $\mathcal{T}$:

- $\Delta(\mathcal{T})$ may be the trivial group consisting of the identity only;
- $\Delta(\mathcal{T})$ may be the group $\langle t[\mathbf{a}] \rangle$ generated by a single non-zero translation $t[\mathbf{a}]$, in which case it is isomorphic to $\mathbb{Z}$, the group of integers under addition;
- $\Delta(\mathcal{T})$ may be the group $\langle t[\mathbf{a}], t[\mathbf{b}] \rangle$ generated by two linearly independent non-zero translations $t[\mathbf{a}]$ and $t[\mathbf{b}]$, in which case it is isomorphic to $\mathbb{Z} \times \mathbb{Z}$.

You may well be able to answer the following exercise without the help of the overlay, but this is a convenient juncture at which to start practising the use of the overlays.

Exercise 1.1

Locate Tiling Card 4 in your *Geometry Envelope*, and also Overlay 1 for Side 1 of this card.

For each of the eight tilings $\mathcal{T}$ depicted on Side 1, decide whether $\Delta(\mathcal{T})$ is trivial, generated by one translation, or generated by two independent translations.

When examining any particular tiling, first place Overlay 1 exactly over Side 1 of the card; this positioning of the overlay represents the identity isometry, which is always a symmetry. Then, concentrating on the particular tiling which you are examining, translate the overlay (without rotating it or turning it over) and see whether you can make the translated copy of the tiling lie exactly over the original: vertices over vertices, edges over edges and tiles over tiles. Each shift of the overlay which achieves this is a translational symmetry of the tiling.

It is important to note that two *independent* translations need not be *orthogonal* (that is, at right angles); for example, for tiling (b), $\Delta(\mathcal{T})$ is generated by two independent translations, and the most obvious pair of translations to take make an angle of $\pi/3$ with each other.

Another important point which arises out of this exercise is that the tiles do *not* all have to be 'facing the same way up' in order for $\Delta(\mathcal{T})$ to be generated by two independent translations. In tilings (f) and (g), for example, the tiles are not all the same way up, despite the fact that $\Delta(\mathcal{T})$ is generated by two independent translations in these cases.

We do not want to keep repeating the awkward phrase 'generated by two independent translations', so we now make some definitions.

Definition 1.1 Point, frieze and wallpaper groups

Let G be a group of isometries of the plane.

- If the translation subgroup is trivial, then G is a **point group**.
- If the translation subgroup is generated by a single non-zero translation, then G is a **frieze group**.
- If the translation subgroup is generated by two independent non-zero translations, then G is a **wallpaper group**.

The name *point group* derives from the fact that there is a point in the plane which remains fixed under the action of a point group. *Frieze* and *wallpaper groups* are so-called because they are the symmetry groups of typical frieze and wallpaper patterns.

Exercise 1.2

Name the point group which is the symmetry group of tiling (a) on Side 1 of Tiling Card 4.

Throughout the remainder of this unit, we shall concentrate on tilings whose symmetry groups are wallpaper groups. Such tilings are themselves given a special name.

Definition 1.2 Periodic tiling

A **periodic tiling** is a tiling whose symmetry group is a wallpaper group.

Exercise 1.3

Which of the tilings depicted on Side 1 of Tiling Card 4 are periodic?

The remainder of the unit is concerned with periodic tilings. There is, however, a possible source of trouble which needs to be guarded against. Consider the tiling in Figure 1.2.

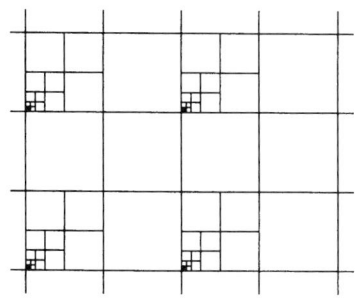

Figure 1.2

Although this tiling is periodic, there are infinitely many tiles in each 'period'! This is because there are certain points where the tiles go into a kind of infinite scrum. Tilings like this are difficult to analyse; they usually disobey theorems which hold in well-behaved cases.

Branko Grünbaum and Geoffrey Shephard have proposed a definition which removes these problems, in the sense that tilings which obey the definition are well-behaved. We shall call it the *little-and-large property*.

See page 121 of *Tilings and Patterns: an introduction* by B. Grünbaum and G.C. Shephard (1989, W.H. Freeman). The property we define here is Property N3.

Definition 1.3 Little-and-large property

A tiling $\mathcal{T}$ has the **little-and-large property** if there are two *positive* real numbers u and U, such that:

(a) some disc of diameter u can be placed entirely within any tile;

(b) any tile can be placed entirely within some disc of diameter U.

The tiling of Figure 1.2 does *not* have the little-and-large property, as it contains arbitrarily small tiles: whatever (positive) value of u is chosen, a tile can be found that is so small that no disc of diameter u can fit within it, so condition (a) fails. On the other hand, although the tiling $\mathcal{T}$ of Figure 1.3 has tiles of various sizes, the disc of diameter u can clearly be moved to fit entirely within any given tile (since, as drawn, it fits within the smallest). Similarly, the disc of diameter U can be moved so that any given tile fits within it (since, as drawn, the biggest fits within it). Therefore $\mathcal{T}$ obeys the little-and-large property.

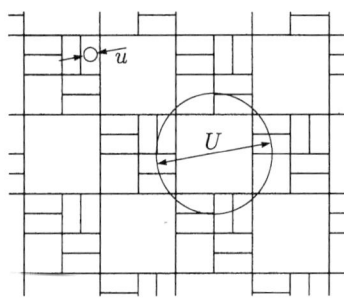

Figure 1.3

Exercise 1.4

Can you think of a tiling which disobeys condition (b) of Definition 1.3?

This is difficult; do not spend too long on it if you cannot see an example fairly quickly.

An important consequence of the little-and-large property is that any tiling with this property is *finitary*, in the following sense.

Definition 1.4 Finitary tiling

A tiling $\mathcal{T}$ is **finitary** if, for any disc D drawn in the plane, D contains, intersects or touches only finitely many parts of $\mathcal{T}$.

Theorem 1.1 Finitary Theorem

Any tiling which has the little-and-large property is finitary.

The proof of this theorem is given in the Appendix and is optional.

In the remainder of this unit, we consider only periodic tilings having the little-and-large property.

1.2 Translational tilings

Of the periodic tilings on Side 1 of Tiling Card 4, tilings (b) and (c) have a special property: *any tile can be mapped to any other tile by a translation belonging to* $\Delta(\mathcal{T})$. This property can be checked using the overlays, as follows. Each periodic tiling is depicted with a small cross in the interior of one of the tiles, both on the card and on Overlay 1. You should verify that, in the case of tilings (b) and (c), but *not* any of the other tilings on Side 1 of Tiling Card 4, the marked tile on the overlay can be made to lie over *any* tile on the underlying tiling, using a translational symmetry.

> **Definition 1.5 Translational tiling**
>
> A tiling $\mathcal{T}$ is **translational** if, for any two tiles S and T of $\mathcal{T}$, there is a translation $t \in \Delta(\mathcal{T})$ such that
>
> $t(S) = T.$

Thus, tilings (b) and (c) on Side 1 of Tiling Card 4 are translational.

In the case of tiling (d), it is worth noting that the tiles are in fact all congruent and all the same way up, despite the fact that the tiling is *not* translational! Any tile can be mapped *individually* to any other using a translation, but in general the required translation does not map *the whole tiling* to itself — that is, it does not belong to the translation group of the tiling.

Exercise 1.5

One of the other tilings on Side 1 of Tiling Card 4 also has the property that all tiles are congruent and the same way up, despite the fact that the tiling is not translational. Which one is it?

Exercise 1.6

Look at Tiling Cards 1 and 3 which depict the Archimedean and Laves tilings respectively. Which of these tilings are translational?

You studied these tilings in *Unit IB1*.

Exercise 1.7

The two tilings on Side 1 of Tiling Card 4 that are translational are tile-uniform; what are their tile types?

There is a further observation that you may possibly have made. Let us temporarily denote tilings (b) and (c) on Side 1 of Tiling Card 4 by $\mathcal{T}_b$ and $\mathcal{T}_c$. Then not only does $\mathcal{T}_b$ have the same *tile type* as $\mathcal{R}_6$ — it actually has the same *structure*: the tiles are put together in the same way. Similarly, $\mathcal{T}_c$ has the same structure as $\mathcal{R}_4$. That is to say, we can find a one–one correspondence between the tiles of $\mathcal{T}_b$ and those of $\mathcal{R}_6$ such that adjacent tiles in $\mathcal{T}_b$ correspond to adjacent tiles in $\mathcal{R}_6$ — and similarly for $\mathcal{T}_c$ and $\mathcal{R}_4$.

In Figure 1.4, we have drawn a portion of $\mathcal{T}_b$ and a portion of $\mathcal{R}_6$, with seven tiles marked 1 to 7 of $\mathcal{T}_b$ corresponding to seven tiles marked 1 to 7 of $\mathcal{R}_6$. Clearly, this can be extended to a one–one correspondence between all the tiles of the two tilings, such that two tiles are adjacent in $\mathcal{T}_b$ if and only if the corresponding tiles are adjacent in $\mathcal{R}_6$.

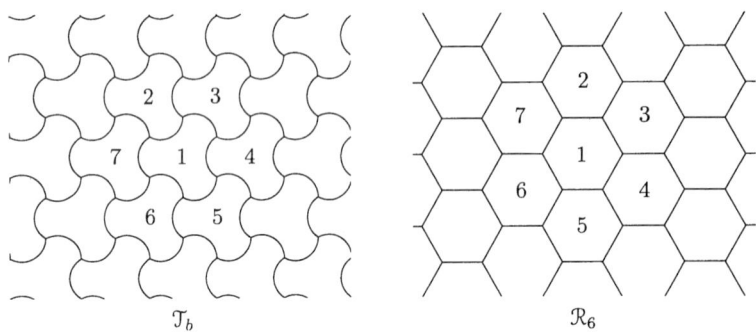

Figure 1.4

Exercise 1.8

Figure 1.5 depicts portions of $\mathcal{T}_c$ and $\mathcal{R}_4$. Label portions of these tilings (as in Figure 1.4) so that the correspondence extends to a one–one correspondence, such that two tiles are adjacent in $\mathcal{T}_c$ if and only if the corresponding tiles are adjacent in $\mathcal{R}_4$.

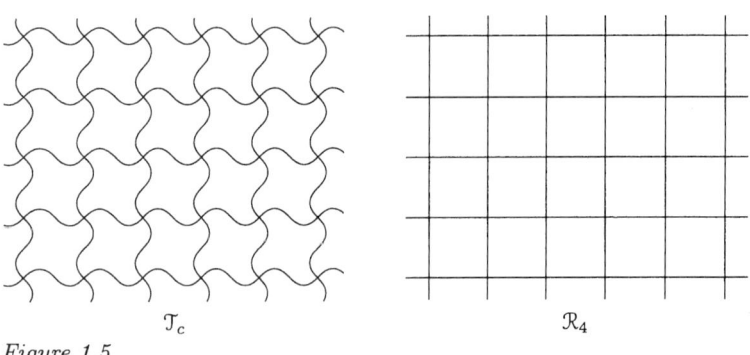

Figure 1.5

Such a correspondence between tiles gives rise to a similar correspondence between edges, and between vertices. Two tilings which can be brought into correspondence in this way are said to be *isomorphic*. This concept is an important one.

> **Definition 1.6 Isomorphic tilings**
>
> Two tilings $\mathcal{S}$ and $\mathcal{T}$ are **isomorphic** if there exists a one–one correspondence between the parts of $\mathcal{S}$ and those of $\mathcal{T}$ such that:
> (a) two parts of the same type in $\mathcal{S}$ are adjacent if and only if the corresponding parts in $\mathcal{T}$ are adjacent;
> (b) two parts of different types in $\mathcal{S}$ are incident if and only if the corresponding parts in $\mathcal{T}$ are incident.
>
> The correspondence itself is called an **isomorphism**.

This definition may seem rather fearsome. Certainly, given two tilings with no pattern, the job of checking whether they are isomorphic would be a tough one. For periodic tilings, however, it is not really difficult. With a little practice, you will find that you can spot isomorphic and non-isomorphic tilings quite quickly by eye.

Exercise 1.9

(a) Classify the four tilings in Figure 1.6 into two pairs of isomorphic tilings.
(b) Label portions of the tilings in Figure 1.6 (as in Figure 1.4) so that the correspondences between the tiles in the isomorphic pairs extend to isomorphisms between the tilings.

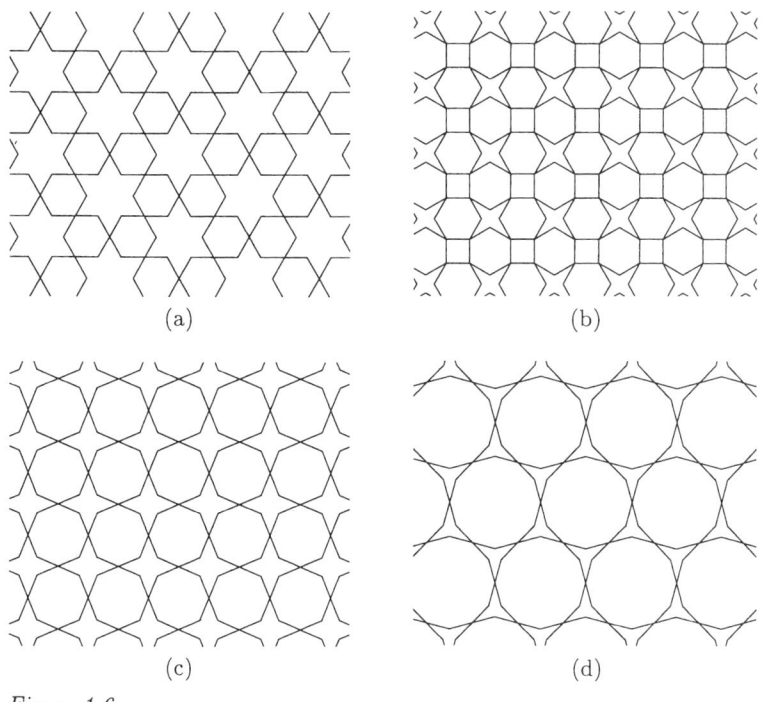

Figure 1.6

In particular, in the case of tilings that are tile-uniform as well as periodic (and have the little-and-large property), you may use the following result.

Theorem 1.2

Let S and T be two tilings having the little-and-large property. Then S and T are isomorphic if they are:

either tile-uniform with the same tile type

or vertex-uniform with the same vertex type.

The proof is rather tough, and does not form part of this course. The tile-uniform part of the proof is given on pages 176–7 of *Tilings and Patterns* (see page 8). The normality condition quoted there is the little-and-large condition plus the requirement that the tile degrees should all be at least 3 (a requirement that does not affect the proof of this theorem).

Exercise 1.10

Use Theorem 1.2 to verify the classification that you made in part (a) of Exercise 1.9.

1.3 Translational orbits

In working with the tiling cards and their overlays, you have not been treating the translation groups as abstract entities. You have been using them actively to discover which tiles can be mapped to other tiles. In other words, you have been considering the *action* of $\Delta(\mathcal{T})$ on $\mathcal{T}$ for various tilings $\mathcal{T}$.

You have met group actions before in this course and in each case you saw that the way in which a group action partitions a set into *orbits* is of great significance. But before we can partition a set into orbits, we need to know exactly what the set in question is! In the case of a tiling $\mathcal{T}$, there are at least four possible sets on which we can consider $\Delta(\mathcal{T})$ to act:

For example, in *Unit IB2* and *Unit GE1*.

- the set of tiles;
- the set of edges;
- the set of vertices;
- the set of parts — that is, the union of the three sets above.

Each of these sets gives rise to a group action of $\Delta(\mathcal{T})$, and questions about group actions on tiles, edges and vertices have all been quite thoroughly investigated in recent years. In this unit, we concentrate on group actions (both by the translation group and by the whole symmetry group) on the *set of tiles* of a tiling, though some use will also be made of the action of the translation group on the set of edges and the set of vertices.

You may recall from *Unit IB2* that two important aspects of the action of a group G on a set X are:

- the orbits into which the set is partitioned by the action (subsets of X);
- the stabilizers of the elements of the set (subgroups of G).

The orbits defined by the action of the translation group are called *translational orbits*.

Definition 1.7 Translational orbits

Let $\mathcal{T}$ be any tiling; then the action of $\Delta(\mathcal{T})$ on the parts of $\mathcal{T}$ partitions the parts into orbits, as follows:

- the tiles are partitioned into orbits known as **translational tile orbits**;
- the edges are partitioned into orbits known as **translational edge orbits**;
- the vertices are partitioned into orbits known as **translational vertex orbits**.

Now look again at tilings (e) to (h) on Side 1 of Tiling Card 4. Each of these tilings is periodic; can you see how many translational tile orbits there are?

When you have given the matter some thought, find Overlay 2 for Side 1 of Tiling Card 4. On it, you will find the tiles divided into translational orbits. For example, tiling (e) has three translational tile orbits, and the tiles therein are marked 1, 2 and 3, respectively, on the overlay. By translating the overlay into different positions over the card as before, you should be able to verify that any tile in a given orbit can be translated to any other using an element of $\Delta(\mathcal{T})$, but that no tile in one orbit can be translated to a tile in a different orbit.

We now need some notation to refer to the numbers of orbits of each kind.

> *Notation Translational orbit numbers*
>
> Let $\mathcal{T}$ be any periodic tiling having the little-and-large property. Then the number of:
> - translational tile orbits of $\mathcal{T}$ is denoted by $n_t(\mathcal{T})$;
> - translational edge orbits of $\mathcal{T}$ is denoted by $n_e(\mathcal{T})$;
> - translational vertex orbits of $\mathcal{T}$ is denoted by $n_v(\mathcal{T})$.

$\mathcal{T}$ must be periodic and must have the little-and-large property in order to ensure that the numbers $n_t(\mathcal{T})$, $n_e(\mathcal{T})$ and $n_v(\mathcal{T})$ are finite. This is a direct consequence of Theorem 1.1, since if $\mathcal{T}$ is periodic, it is possible to draw a disc large enough to contain a complete 'period'; this disc must therefore contain, touch or intersect an element of every translational tile, edge and vertex orbit.

Exercise 1.11

Determine $n_t(\mathcal{T})$ for each of the tilings on Tiling Cards 1 and 3, and then check your result by using Overlays 2 for both sides of the cards.

Finding translational edge and vertex orbits requires just the same technique as finding translational tile orbits. Although this task is admittedly rather time consuming, it is an excellent way to become familiar with group actions on tilings, and you may wish to do this for some of the tilings on Tiling Card 1, as well as for tilings (e) to (h) on Side 1 of Tiling Card 4. (You can always leave some of them for revision purposes.)

Your results should be as shown in Table 1.1.

$\mathcal{T}$	$n_t(\mathcal{T})$	$n_e(\mathcal{T})$	$n_v(\mathcal{T})$
Tiling Card 1:			
$(3,3,3,3,3,3)$	2	3	1
$(4,4,4,4)$	1	2	1
$(6,6,6)$	1	3	2
$(3,3,3,3,6)$	9	15	6
$(3,3,3,4,4)$	3	5	2
$(3,3,4,3,4)$	6	10	4
$(3,4,6,4)$	6	12	6
$(3,6,3,6)$	3	6	3
$(3,12,12)$	3	9	6
$(4,6,12)$	6	18	12
$(4,8,8)$	2	6	4
Tiling Card 4:			
(e)	3	8	5
(f)	2	3	1
(g)	2	4	2
(h)	18	30	12

Table 1.1

Exercise 1.12

Can you see the simple relationship between $n_t(\mathcal{T})$, $n_e(\mathcal{T})$ and $n_v(\mathcal{T})$, as revealed in Table 1.1?

2 THE EULER EQUATION

In the solution to Exercise 1.12, you will have come across the simple equation

$$n_t(\mathcal{T}) - n_e(\mathcal{T}) + n_v(\mathcal{T}) = 0. \tag{2.1}$$

Leonhard Euler discovered a similar relationship between the numbers of faces, edges and vertices of a polyhedron P (denoted by $n_f(P)$, $n_e(P)$ and $n_v(P)$), respectively, namely

$$n_f(P) - n_e(P) + n_v(P) = 2.$$

He would not have recognized the concepts of tile, vertex and edge *orbits*, since group theory had not been invented in his day, but nevertheless we shall honour his name by calling Equation 2.1 the *Euler Equation*. The purpose of this short section is to prove the validity of this equation.

The proof is by induction on $n_t(\mathcal{T})$, the number of translational tile orbits. In order to get started, therefore, we need to verify the Euler Equation for tilings having $n_t(\mathcal{T}) = 1$ — that is, for translational tilings.

Now the translational tilings which you encountered in Section 1 are all isomorphic to either $\mathcal{R}_4$ or $\mathcal{R}_6$, and you may well have wondered at the time whether this is true for *any* translational tiling. It is indeed, and a discussion of this result, together with an enumeration of the translational orbits in each case, follows.

Leonhard Euler (1707–83) was a remarkably prolific mathematician, who made fundamental contributions to several mathematical areas including topology, combinatorics and number theory.

You may have met this as $f - e + v = 2$.

2.1 Translational Tiling Theorem

Theorem 2.1 Translational Tiling Theorem

Let $\mathcal{T}$ be a translational tiling. Then $\mathcal{T}$ is tile-uniform, and is *either* of tile type $[4, 4, 4, 4]$, with

$$n_t(\mathcal{T}) = 1, \ n_e(\mathcal{T}) = 2, \ n_v(\mathcal{T}) = 1$$

or of tile type $[3, 3, 3, 3, 3, 3]$, with

$$n_t(\mathcal{T}) = 1, \ n_e(\mathcal{T}) = 3, \ n_v(\mathcal{T}) = 2.$$

Sketch of proof

We now sketch an informal argument which illustrates why Theorem 2.1 is true.

We begin by noting that if T and T' are adjacent tiles of a translational tiling $\mathcal{T}$, then the translational symmetry $t[\mathbf{a}]$ which maps T to T' can be repeated indefinitely, as can its inverse. That is to say, we can consider the subgroup

$$G = \langle t[\mathbf{a}] \rangle = \{(t[\mathbf{a}])^n : n \in \mathbb{Z}\}$$

of $\Delta(\mathcal{T})$.

The set of all images of T under the translations $(t[\mathbf{a}])^n$ constitutes an infinite strip of tiles (see Figure 2.1). Call this strip $\mathcal{S}$.

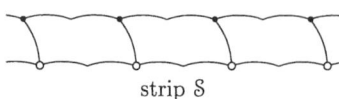

strip $\mathcal{S}$

Figure 2.1

Now consider the action of G on the set $\mathcal{T}$ of all tiles. As with any group action, it partitions $\mathcal{T}$ into orbits. The orbit containing T is the strip $\mathcal{S}$; the other orbits lie parallel to $\mathcal{S}$, and cover the whole of $\mathcal{T}$. Each orbit is an infinite strip, like $\mathcal{S}$, and, because the orbits partition the whole of $\mathcal{T}$, they must fit flush with each other, without leaving any gaps. There are two ways in which they can do this, illustrated in Figures 2.2 and 2.3. (Alternate strips are shown shaded and unshaded.)

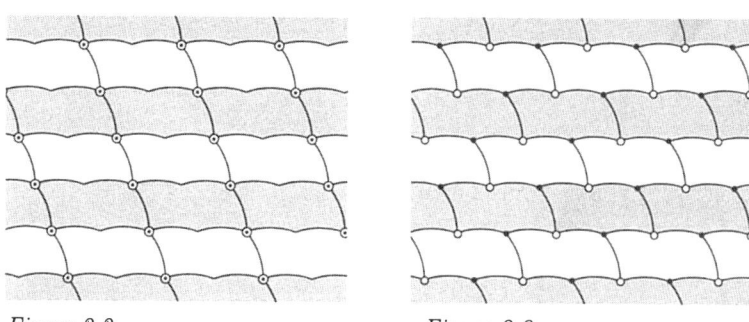

Figure 2.2   *Figure 2.3*

To understand why there are just these two possibilities, look back to Figure 2.1. The vertices on the upper side of $\mathcal{S}$ are drawn as black dots, and those on the lower side are drawn as white dots. The black set and the white set each form a vertex orbit under the action of G; hence all the black dots are in the same orbit under $\Delta(\mathcal{T})$, just as all the white dots are in the same orbit.

But when we use a translation *not* in G to map $\mathcal{S}$ to the strip immediately above, one of two things can happen. Either the black dots are superimposed on the white dots, forming just one orbit under $\Delta(\mathcal{T})$ (see Figure 2.2), or the black and white dots remain in two separate orbits under $\Delta(\mathcal{T})$ (see Figure 2.3).

Figure 2.2 clearly gives us a tiling of tile type $[4,4,4,4]$, each tile being adjacent to two tiles in its own strip, one tile in the strip above and one in the strip below. As $\mathcal{T}$ is translational, $n_t(\mathcal{T}) = 1$ and, because the black and white dots have merged into one orbit, $n_v(\mathcal{T}) = 1$. Finally, each edge of $\mathcal{T}$ is either an edge which divides one tile from another *within* a strip, or an edge dividing two tiles in *different* strips. Thus, $n_e(\mathcal{T}) = 2$.

On the other hand, Figure 2.3 gives us a tiling of tile type $[3,3,3,3,3,3]$, because each tile is adjacent to two tiles in its own strip, two in the strip above and two in the strip below. As $\mathcal{T}$ is translational, $n_t(\mathcal{T}) = 1$, and because the black and white dots form distinct translational orbits, $n_v(\mathcal{T}) = 2$. Finally, each edge of $\mathcal{T}$ is either an edge dividing one tile from another within a strip, or in one of *two* edge orbits dividing tiles in *different* strips. Thus, $n_e(\mathcal{T}) = 3$. ∎

One of these edge orbits has a black dot on the left and a white on the right; the other has them the other way round.

Although we have used concepts such as group actions and orbits in the above argument, it is nevertheless not a full proof; at various points we have simply asserted that things 'must be' of one form or another. Justifying all this in complete rigour would (as so often in this course!) involve a detour into topology which we are not in a position to make.

Exercise 2.1

Can you think of a simple argument to prove that no tiling $\mathcal{T}$ can have $n_e(\mathcal{T}) = 1$?

2.2 The Euler Equation

> **Theorem 2.2 Euler Equation**
>
> The Euler Equation
> $$n_t(\mathcal{T}) - n_e(\mathcal{T}) + n_v(\mathcal{T}) = 0 \qquad (2.2)$$
> holds for any periodic tiling $\mathcal{T}$ which has the little-and-large property.

Proof

As we mentioned at the beginning of this section, the proof is by induction on $n_t(\mathcal{T})$.

Step 1

If $n_t(\mathcal{T}) = 1$, then $\mathcal{T}$ is translational. Thus, by Theorem 2.1,
either $\quad n_e(\mathcal{T}) = 2 \quad$ and $\quad n_v(\mathcal{T}) = 1$.
or $\quad n_e(\mathcal{T}) = 3 \quad$ and $\quad n_v(\mathcal{T}) = 2$
In each case, the Euler Equation is satisfied.

Step 2

We *assume* that the Euler Equation holds for any periodic tiling, which has the little-and-large property, and for which

$$n_t(\mathcal{T}) = k,$$

where k is some positive integer.

We let $\mathcal{T}$ be a periodic tiling, obeying the little-and-large property, and such that

$$n_t(\mathcal{T}) = k + 1.$$

Figure 2.4

Let T_1 and T_2 be tiles which are from two different translational orbits, and which are adjacent along a common edge, E (see Figure 2.4). (It must be possible to find such a pair, since if T_1 were adjacent only to tiles from the *same* translational orbit, then all the tiles adjacent to T_1 would have the same property, and we would find that we could never get away from tiles in the same translation orbit as T_1. Thus $n_t(\mathcal{T})$ would equal 1, contradicting our assumption that $n_t(\mathcal{T}) = k + 1$.)

Now remove E. This amalgamates T_1 and T_2 into a single tile, say T_3, of a new tiling. Remove all the edges in the same translational edge orbit as E. Then we again have a periodic tiling $\mathcal{S}$. We certainly have not added any tiles, edges or vertices, so $\mathcal{S}$ must obey the little-and-large property. The tiles T_1 and T_2 of $\mathcal{T}$ have merged into a single tile T_3, and all their translated images have also merged into single tiles. Therefore, the translational tile orbits of $\mathcal{T}$ containing T_1 and T_2 have merged into a single tile orbit of $\mathcal{S}$, containing T_3. Thus,

$$n_t(\mathcal{S}) = n_t(\mathcal{T}) - 1 \qquad (2.3)$$
$$= k.$$

By our assumption of Step 2, therefore,

$$n_t(\mathcal{S}) - n_e(\mathcal{S}) + n_v(\mathcal{S}) = 0. \qquad (2.4)$$

Now let us compare $n_e(\mathcal{S})$ with $n_e(\mathcal{T})$, and $n_v(\mathcal{S})$ with $n_v(\mathcal{T})$. The simplest case to consider is where the vertices incident with E in $\mathcal{T}$ are of degree 4 or more (see Figure 2.5).

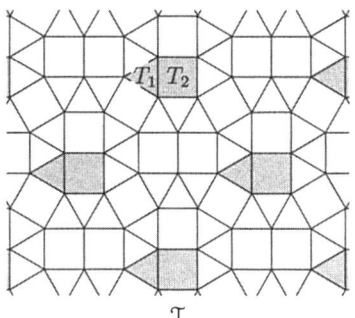 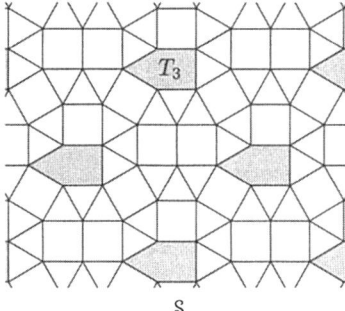

Figure 2.5

Clearly no vertices disappear, nor do any vertex orbits, while exactly one edge orbit (the orbit containing E) disappears on converting $\mathcal{T}$ to $\mathcal{S}$. Thus,

$$n_e(\mathcal{S}) = n_e(\mathcal{T}) - 1, \qquad (2.5)$$
$$n_v(\mathcal{S}) = n_v(\mathcal{T}). \qquad (2.6)$$

Now, substituting the right-hand sides of Equations 2.3, 2.5 and 2.6 into Equation 2.4, we obtain

$$(n_t(\mathcal{T}) - 1) - (n_e(\mathcal{T}) - 1) + n_v(\mathcal{T}) = 0;$$
$$n_t(\mathcal{T}) - n_e(\mathcal{T}) + n_v(\mathcal{T}) = 0;$$

and so the Euler Equation is valid for $\mathcal{T}$.

The next simplest case is where one of the vertices incident with E in $\mathcal{T}$ is of degree 4 or more, and the other (V, say) is of degree 3 (see Figure 2.6).

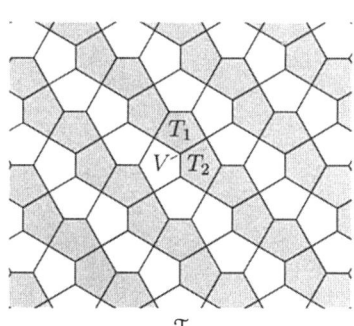 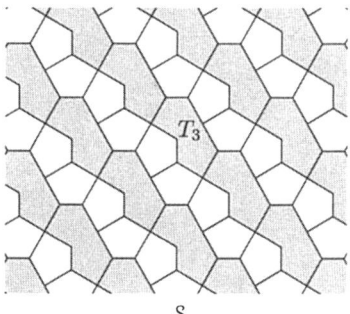

Figure 2.6

In this case, when we remove E, the vertex V of order 3 disappears as well! This is also true for every vertex in the corresponding translational orbit. However, the two translational edge orbits corresponding to the other two edges incident with V coalesce into one. Thus,

$$n_e(\mathcal{S}) = n_e(\mathcal{T}) - 2, \qquad (2.7)$$
$$n_v(\mathcal{S}) = n_v(\mathcal{T}) - 1, \qquad (2.8)$$

and substituting these and Equation 2.3 into Equation 2.4, we again obtain

$$n_t(\mathcal{T}) - n_e(\mathcal{T}) + n_v(\mathcal{T}) = 0$$

so that the Euler Equation is again valid for $\mathcal{T}$.

There is one more case to consider. We invite you to explore this in the exercise which follows. You should again find that the Euler Equation for $\mathcal{T}$ is satisfied. □

Exercise 2.2

Which case remains to be considered? Show that the Euler Equation for $\mathcal{T}$ is also satisfied in this case.

Proof of Theorem 2.2 *continued*

This completes the inductive step, and the proof is finished. ∎

3 ORBIT DIAGRAMS

The aim of this section is to show you how to produce, from a periodic tiling that may be awkward to draw, a set of three *finite* diagrams that record the incidence relations between the translational tile, edge and vertex orbits. These diagrams turn out to be very useful in the analysis of periodic tilings.

In order to produce these diagrams, we need to be able to identify the orbits. You have seen the tile orbits for Tiling Card 3 labelled by means of Overlays 2 for each side. If you now look at Overlays 3 and 4 for both sides of Tiling Card 3, you will see that Overlays 3 label the edge orbits and Overlays 4 label the vertex orbits.

3.1 Tile–edge diagram

Now look in detail at the bottom right-hand tiling on Side 1 of Tiling Card 3, the Laves tiling [3, 3, 3, 4, 4]. It is a little awkward to have Overlays 2, 3 and 4 all on the card, so just place Overlay 3 on the card, giving you the edge orbits, and remember that the upwards pointing tiles form tile orbit 1 and the downwards pointing tiles form tile orbit 2.

You will see that each tile in tile orbit 1 is incident with one edge from each of the edge orbits 1, 4 and 5, and two edges from edge orbit 2. Each tile in tile orbit 2 is incident with one edge from each of the edge orbits 1, 4 and 5 and two edges from edge orbit 3. We record these facts in a diagram, which we call a *tile–edge diagram* (see Figure 3.1).

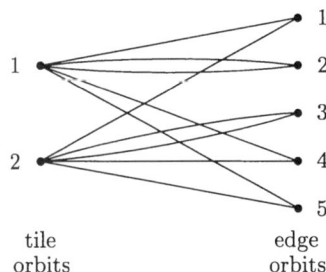

Figure 3.1   Tile–edge diagram for Laves tiling [3, 3, 3, 4, 4].

Now turn over to Side 2 of Tiling Card 3, and again look at the bottom right-hand corner, at the Laves tiling [4, 8, 8]. First use Overlay 2, and note that the tile orbit labels 1, 2, 3, 4 occur clockwise round each vertex of order 4. Then use Overlay 3, and note which edge orbits are incident with each of the four vertex orbits. You should obtain the diagram in Figure 3.2.

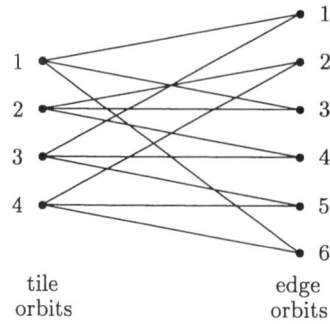

Figure 3.2   Tile–edge diagram for Laves tiling [4, 8, 8].

Exercise 3.1

Draw the tile–edge diagram for the third tiling on Side 2 of Tiling Card 3, namely the Laves tiling [3, 6, 3, 6], and for the second tiling on Side 1, namely the Laves tiling [4, 4, 4, 4].

You may have noticed two facts about these diagrams:
- the number of lines coming out of the dot for a tile orbit is equal to the degree of that tile;
- the number of lines going into the dot for each edge orbit is two.

The first of these facts is not surprising — we drew a line corresponding to each edge of the tile for each tile orbit. The second is a little more surprising, until we remember that each edge of a tiling has two sides! Each line in the tile–edge diagram in fact corresponds to *one side* of a typical edge in one of the translational edge orbits. If both sides of an edge are in tiles belonging to the same tile orbit, then we get two lines between the dots representing that edge orbit and that tile orbit.

Exercise 3.2

Draw the tile–edge diagram for the Laves tiling [3, 3, 3, 3, 3, 3].

3.2 Vertex–edge diagram

The *vertex–edge diagram* for a periodic tiling is similar in construction to the tile–edge diagram. This time, we draw a dot on the left for each vertex orbit, and a dot on the right for each edge orbit. Then we draw lines to represent the incidence relations, once again drawing two lines from a vertex dot to an edge dot if a vertex in a vertex orbit is incident with two edges in a particular edge orbit.

Using Overlay 4 for Side 1 of Tiling Card 3, and again considering the Laves tiling [3, 3, 3, 4, 4], we obtain the vertex–edge diagram in Figure 3.3.

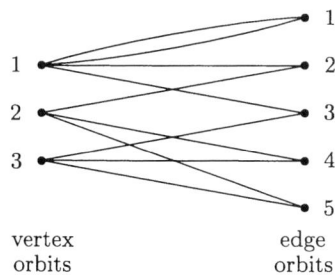

Figure 3.3   Vertex–edge diagram for Laves tiling [3, 3, 3, 4, 4].

Again, we note the following facts:
- the number of lines coming out of the dot for a vertex orbit is equal to the degree of that vertex;
- the number of lines going into the dot for each edge orbit is two.

For the vertex–edge diagram, each line corresponds to *one end* of a typical edge in one of the translational edge orbits. Thus, this time, two lines between a pair of dots in the diagram mean that, for a typical edge in an edge orbit, the vertices at *either end* of that edge are in the same vertex orbit.

Exercise 3.3

Draw the vertex–edge diagrams for the Laves tilings [3, 6, 3, 6], [4, 4, 4, 4] and [3, 3, 3, 3, 3, 3].

3.3 Tile–vertex diagram

The *tile–vertex diagram* is exactly what you would expect: we draw a dot on the left for each tile orbit, and a dot on the right for each vertex orbit. Then we draw lines to represent the incidence relations. In this case, though, we may have to draw more than two lines between certain pairs of dots. For example, in the Laves tiling [6, 6, 6], there are two tile orbits and just one vertex orbit, but each tile is incident with three vertices from the one vertex orbit. Correspondingly, each vertex is incident with three tiles from each orbit, so we obtain the tile–vertex diagram with 3-fold lines.

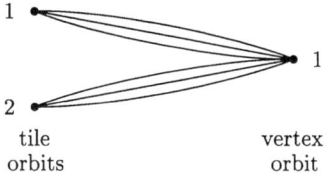

Figure 3.4 Tile–vertex diagram for Laves tiling [6, 6, 6].

Exercise 3.4

Draw the tile–vertex diagrams for the Laves tilings [3, 3, 3, 4, 4], [3, 6, 3, 6] and [4, 4, 4, 4].

As with the tile–edge and vertex–edge diagrams, two facts about the tile–vertex diagram are worth noting:
- the number of lines coming out of the dot for a tile orbit is equal to the degree of that tile;
- the number of lines going into the dot for a vertex orbit is equal to the degree of that vertex.

For the tile–vertex diagram, each line corresponds to an *angle* made at a typical vertex in one of the vertex orbits. If, at a typical vertex, there are k different angles facing into tiles from the same tile orbit, then each of those tiles must have k angles from that orbit facing into it, so we get a k-fold line.

3.4 Using orbit diagrams

It is often possible to use orbit diagrams without actually drawing them! That is to say, the fact that it would be *possible* to draw an orbit diagram can often allow us to infer some information about a tiling.

Example 3.1

Let $\mathcal{T}$ be the Laves tiling [3, 3, 3, 4, 4]. Given that $n_t(\mathcal{T}) = 2$, use the properties of the tile–edge diagram to calculate $n_e(\mathcal{T})$.

Solution

The tile–edge diagram has two dots representing tile orbits. As each tile is of degree 5, there are $2 \times 5 = 10$ lines going from the tile dots to the edge dots. But each edge dot has two lines going into it, so there are $\frac{10}{2} = 5$ edge dots. Thus,
$$n_e(\mathcal{T}) = 5.$$ ♦

The vertex–edge diagram can be used in exactly the same way; we leave this as an exercise.

Exercise 3.5

Let $\mathcal{T}$ be the Archimedean tiling of vertex type (3, 4, 6, 4), which has six translational vertex orbits. Use the properties of the vertex–edge diagram to calculate $n_e(\mathcal{T})$.

Example 3.2

Tiling (h) on Side 1 of Tiling Card 4 has six translational tile orbits consisting of squares and twelve consisting of triangles. How many translational edge orbits does it have?

Solution

The tile–edge diagram has six dots representing square tile orbits, each with four lines coming out, and twelve dots representing triangular tile orbits, each with three lines coming out. Thus there are $24 + 36 = 60$ lines going from the tile dots to the edge dots. But (as before) each edge dot has two lines going into it, so there are $\frac{60}{2} = 30$ translational edge orbits. ♦

Exercise 3.6

Let $\mathcal{T}$ be tiling (e) on Side 1 of Tiling Card 4. Find the degree of a typical tile in each of the three translational tile orbits, and hence find $n_e(\mathcal{T})$.

We can 'automate' the arguments we have used in the last two examples and exercises.

Theorem 3.1 Edge Orbit Theorem

Let $\mathcal{T}$ be a tiling having n_3 translational orbits of tiles of degree 3, n_4 of tiles of degree 4, ..., n_k of tiles of degree k, and so on. Then

$$n_e(\mathcal{T}) = \tfrac{1}{2}(3n_3 + 4n_4 + \cdots + kn_k + \cdots). \tag{3.1}$$

The same result holds if the n_k represent numbers of translational orbits of vertices of various degrees.

Proof

In the tile–edge diagram (or vertex–edge diagram, as the case may be) there are $3n_3$ lines coming from the dots representing orbits of tiles (or vertices) of degree 3, $4n_4$ from those representing tiles (or vertices) of order 4, and so on. Thus there are $3n_3 + 4n_4 + \cdots + kn_k + \cdots$ lines altogether. Now each edge orbit dot has two lines going into it, and so

$$n_e(\mathcal{T}) = \tfrac{1}{2}(3n_3 + 4n_4 + \cdots + kn_k + \cdots). \qquad \blacksquare$$

When using the tile–vertex diagram we need to take a little more care, as both the tiles and the vertices may vary in their degrees. The trick is to give separate consideration to the lines going into orbit dots representing vertices of different degrees.

Example 3.3

The Laves tiling $[3, 3, 4, 3, 4]$ has four translational tile orbits. How many translational vertex orbits does it have?

Solution

The tile–vertex diagram has four dots representing translational tile orbits. Now each tile has three vertices of degree 3, so each dot representing a tile orbit has three lines going from it *to dots representing orbits of vertices of degree 3*. Thus, there are $4 \times 3 = 12$ lines in the diagram *going into dots representing translational orbits of vertices of degree 3*. Therefore the number of such dots is $\frac{12}{3} = 4$.

Also, each tile has two vertices of degree 4, so each dot representing a tile orbit has two lines going from it *to dots representing orbits of vertices of degree 4*. Thus, there are $4 \times 2 = 8$ lines in the diagram *going into dots representing translational orbits of vertices of degree 4*. Therefore the number of such dots is $\frac{8}{4} = 2$.

Finally, therefore, the total number of translational vertex orbits is $4 + 2 = 6$. ♦

To put it another way, of the twenty lines in the diagram, those that go to dots representing orbits of vertices of degree 3 each contribute 'one third of a vertex orbit', while those that go to dots representing orbits of vertices of degree 4 each contribute 'one quarter of a vertex orbit'. Thus, the tile type $[3,3,4,3,4]$ can be read as '$\frac{1}{3}$ of a vertex orbit, $\frac{1}{3}$ of a vertex orbit, $\frac{1}{4}$ of a vertex orbit, $\frac{1}{3}$ of a vertex orbit, $\frac{1}{4}$ of a vertex orbit'. Each dot representing a tile orbit therefore produces the lines that contribute to $\left(\frac{1}{3} + \frac{1}{3} + \frac{1}{4} + \frac{1}{3} + \frac{1}{4}\right)$ vertex orbits, and so the total number of translational vertex orbits is $4\left(\frac{1}{3} + \frac{1}{3} + \frac{1}{4} + \frac{1}{3} + \frac{1}{4}\right)$. This illustrates the following theorem.

> **Theorem 3.2 Vertex Orbit Theorem**
>
> Let $\mathcal{T}$ be a periodic tile-uniform tiling of tile type $[x_1, x_2, \ldots, x_q]$. Then
>
> $$n_v(\mathcal{T}) = n_t(\mathcal{T}) \left(\frac{1}{x_1} + \frac{1}{x_2} + \cdots + \frac{1}{x_q} \right). \tag{3.2}$$

Proof

In the tile–vertex diagram, each dot representing a tile orbit has coming out of it:

> a line going to an orbit of vertices of degree x_1;
> a line going to an orbit of vertices of degree x_2;
> $\vdots$
> a line going to an orbit of vertices of degree x_q.

Now of the lines going to the orbits of vertices of degree x_1, each makes $\frac{1}{x_1}$ of the total contribution to the lines going into these orbits. Thus the symbol x_1 'generates' $n_v(\mathcal{T}) \times \left(\frac{1}{x_1}\right)$ vertex orbits. The same argument applies to the symbols $x_2, \ldots, x_q$ in the tile type, and so, in total,

$$n_v(\mathcal{T}) = n_t(\mathcal{T}) \left(\frac{1}{x_1} + \frac{1}{x_2} + \cdots + \frac{1}{x_q} \right),$$

as required. ∎

Exercise 3.7

The Laves tiling $[3, 4, 6, 4]$ has six translational tile orbits. How many translational vertex orbits does it have?

Exercise 3.8

The tile–vertex diagram can be drawn with vertices on the left and tiles on the right. Enunciate a theorem which can be derived from the properties of the diagram when viewed in this way.

4 TRANSITIVE TILINGS

4.1 Transitive group actions

So far in this course, most of the group actions you have met have partitioned the set into *several* orbits. In the remainder of this unit, however, we shall consider group actions where all the elements of the set are in a *single* orbit. The definition of a translational tiling (page 9), for example, can be rephrased as follows: *a tiling $\mathcal{T}$ is translational if there is just one translational tile orbit.*

This property of a group action has a name: *transitivity*.

Definition 4.1 Transitive group action

Let any group G act on any set X. Then the group action is **transitive** if all the elements of X belong to a single orbit; that is, given any two elements x, y of X there is an element g of G such that

$$g \wedge x = y.$$

The concept of transitivity makes sense in an algebraic as well as a geometric context, and the following exercise should help you to tie the two streams of the course together.

Exercise 4.1

In *Unit IB2* you met the group action $g \wedge x = gxg^{-1}$ in which the group is D_6 and the set X is the set of elements of D_6.

There is another group action in which D_6 is both the group and the set, namely $g \wedge x = gx$.

Which one of these actions is transitive?

We now return to the topic of group actions on tilings, via another exercise.

Exercise 4.2

For which of the Archimedean tilings does the translation group act transitively on:

(a) the tiles?
(b) the edges?
(c) the vertices?

The translation group of a tiling $\mathcal{T}$ has a rather simple structure, and it is interesting to move on to the more richly structured symmetry group, $\Gamma(\mathcal{T})$. We concentrate on this group in the last two sections of this unit.

Exercise 4.3

Of the tilings on Tiling Cards 1 and 3, and Side 1 of Tiling Card 4, which have symmetry groups that act transitively on:

(a) the tiles?
(b) the edges?
(c) the vertices?

In a series of papers published in the late 1970s, Grünbaum and Shephard carried out a full analysis of tilings having each of these kinds of transitivity. There is not space in this unit to pursue all three analyses, and so we concentrate on transitivity on the tiles.

> **Definition 4.2 Tile-transitive**
>
> A tiling $\mathcal{T}$ is said to be **tile-transitive** (or simply **transitive**) if $\Gamma(\mathcal{T})$ acts transitively on the set of tiles of $\mathcal{T}$.

The term used by Grünbaum and Shephard is *isohedral*.

Exercise 4.3 may have suggested to you that, at least for the tilings on Tiling Cards 1, 3 and 4, the tile-transitive tilings are isomorphic to Laves tilings. This is in fact generally true, and proving it is a good way to become familiar with the transitive tilings and their symmetry groups. The main aim of this section is to prove this result. The first step is to prove that transitive tilings are periodic; this is done in the next subsection.

4.2 Transitivity and periodicity

By definition, any transitive tiling has enough symmetries to map any tile to any other. Those symmetries that are *not* translations must be rotations, reflections or glide reflections; this is a direct consequence of the classification of isometries which we performed in Section 5 of *Unit IB1*.

Essentially, the proof that all transitive tilings are periodic consists in showing that we can compose rotational or glide reflection symmetries to produce translational symmetries. In particular, we need to know that:

- given three rotational symmetries through the same angles but about three non-collinear rotation centres, we can produce two independent translational symmetries;
- given two glide reflection symmetries whose axes are parallel but distinct, we can produce two independent translational symmetries.

The steps required to do this give useful practice in the manipulation of isometries, and we ask you to work through them in the four exercises which follow.

Exercise 4.4

Figure 4.1 depicts a tiling whose vertices are at the integer points of $\mathbb{R}^2$, and whose symmetry group contains rotations through $\pi/2$ about each vertex. If P, Q and R are $(0,0)$, $(1,0)$ and $(1,1)$ respectively, and θ_P, θ_Q and θ_R are the anticlockwise rotations through $\pi/2$ about these points, namely $r[(0,0), \pi/2], r[(1,0), \pi/2]$ and $r[(1,1), \pi/2]$, find the symmetries $\theta_Q^{-1} \theta_P$ and $\theta_R^{-1} \theta_P$.

Figure 4.1

Exercise 4.5

(a) Let θ be any angle which is not a multiple of 2π, let P, Q and R be any three non-collinear points, and let θ_P, θ_Q and θ_R be anticlockwise rotations through θ about P, Q and R. Denote the vectors corresponding to P, Q and R by $\mathbf{p}$, $\mathbf{q}$ and $\mathbf{r}$, respectively, and let $\mathbf{A}$ be the matrix

$$\mathbf{A} = \begin{bmatrix} 1 - \cos\theta & -\sin\theta \\ \sin\theta & 1 - \cos\theta \end{bmatrix}.$$

Show that

$$\theta_Q^{-1}\theta_P = t[\mathbf{a}], \text{ where } \mathbf{a} = \mathbf{A}(\mathbf{q} - \mathbf{p}),$$

and that

$$\theta_R^{-1}\theta_P = t[\mathbf{b}], \text{ where } \mathbf{b} = \mathbf{A}(\mathbf{r} - \mathbf{p}).$$

(b) Show that $\mathbf{a}$ and $\mathbf{b}$ are linearly independent.

Hint Find $\det \mathbf{A}$ and use the fact that $\mathbf{q} - \mathbf{p}$ and $\mathbf{r} - \mathbf{p}$ are linearly independent.

Exercise 4.6

Figure 4.2 depicts a tiling $\mathcal{T}$ which has glide reflection symmetries whose axes are parallel to the x-axis and intersect the y-axis at integer points.

Figure 4.2

In particular, in the notation of the Isometry Toolkit, the isometries

$$g_1 = q[(1,0),(0,1),0], \quad g_2 = q[(1,0),(0,2),0]$$

are symmetries of $\mathcal{T}$. Find the symmetries g_1^2 and $g_2 g_1$.

Exercise 4.7

Let $g_1 = q(\mathbf{g}, \mathbf{c}, \theta)$ and $g_2 = q(\lambda \mathbf{g}, \mu \mathbf{c}, \theta)$ be glide reflections through two distinct parallel glide reflection axes, where $\mathbf{g}$ and $\mathbf{c}$ are non-zero vectors parallel and perpendicular to these axes, and $\mu \neq 1$. Show that g_1^2 and $g_2 g_1$ are translations in linearly independent directions.

Theorem 4.1

Every transitive tiling is periodic.

Proof

Let $\mathcal{T}$ be a transitive tiling.

Our objective is to prove that $\Delta(\mathcal{T})$ contains translations in two independent directions. It turns out to be useful to consider separately the cases where:

1 $\Gamma(\mathcal{T})$ contains non-trivial rotations;
2 $\Gamma(\mathcal{T})$ contains no non-trivial rotations, but contains indirect symmetries;
3 $\Gamma(\mathcal{T})$ contains neither indirect symmetries nor non-trivial rotations.

These three cases clearly cover all possibilities.

Case 1 $\Gamma(\mathcal{T})$ *contains non-trivial rotations.*

Let P be a rotation centre of some non-trivial rotation in $\Gamma(\mathcal{T})$. The set of all rotations in $\Gamma(\mathcal{T})$ having P as a rotation centre is a subgroup of $\Gamma(\mathcal{T})$; call it Γ_P.

Now P is either in the interior or on the boundary of some tile T. As $\mathcal{T}$ is transitive, it follows that *every* tile has some point in its interior or on its boundary which is an image of P under some element of $\Gamma(\mathcal{T})$. Let Q and R be two such points, chosen such that P, Q, R are non-collinear (see Figure 4.3).

Figure 4.3

Let θ_P be a non-trivial rotation belonging to Γ_P. Since Q and R are images of P under certain symmetries of $\mathcal{T}$, it follows that Γ_Q and Γ_R are isomorphic to Γ_P, and so there must be rotations θ_Q and θ_R, about Q and R respectively, through the same angle as θ_P.

Then $\theta_Q^{-1}\theta_P$ and $\theta_R^{-1}\theta_P$ each belong to $\Gamma(\mathcal{T})$, and (by Exercise 4.5) are a pair of linearly independent translations. Thus, $\Delta(\mathcal{T})$ contains translations in two independent directions.

Case 2 $\Gamma(\mathcal{T})$ *contains no non-trivial rotations, but contains indirect symmetries.*

Each indirect symmetry is a reflection or a glide reflection by Theorem 5.1 of *Unit IB1*. Now all their axes must be parallel, since the composite of two such symmetries having non-parallel axes would be a non-trivial rotation. Moreover, since $\mathcal{T}$ is transitive, the tiling looks the same from the point of view of any one tile as from that of any other. Therefore, if there is one axis of reflection or glide reflection symmetry, there must be infinitely many, since there must be at least one such axis passing through or touching every tile of $\mathcal{T}$.

If these symmetries were all reflections, then there would be no way to map a tile onto another tile further along the same reflection axis; so, because $\mathcal{T}$ is transitive, there must be glide reflections in $\Gamma(\mathcal{T})$. Let g_1 and g_2 be glide reflections through two distinct parallel axes; then g_1^2 and $g_2 g_1$ each belong to $\Gamma(\mathcal{T})$, and (by Exercise 4.7) are a pair of linearly independent translations. Therefore, $\Delta(\mathcal{T})$ contains translations in two independent directions.

Case 3 $\Gamma(\mathcal{T})$ *contains neither indirect symmetries nor non-trivial rotations.*

In this case, the *only* non-trivial symmetries are translations; but since $\Gamma(\mathcal{T}) = \Delta(\mathcal{T})$ must contain enough symmetries to map any tile to any other tile, the translations cannot all be in the same direction. Thus, $\Delta(\mathcal{T})$ contains translations in two independent directions. ∎

4.3 The Transitive Tiling Theorem

We have just seen that every transitive tiling is periodic. This is hardly a surprising result, and if we are to obtain a reasonably informative classification of transitive tilings, we need a much stronger result than this. We can indeed obtain a stronger result, by using the Euler Equation and the theorems which we derived in Subsection 3.4.

> *Theorem 4.2 Transitive Tiling Theorem*
> Every transitive tiling is isomorphic to one of the eleven Laves tilings.

Proof

As with the proof of the Archimedean Tiling Theorem in *Unit IB1*, there is a reasonably straightforward part, which produces a list of possible tile types, and then a tedious part in which we exclude some of these types on the grounds that, when we start trying to construct such a tiling, we run into an incompatibility. Just as in *Unit IB1*, we shall present the straightforward part but not the tedious part! Let $\mathcal{T}$ be a transitive tiling. It must be tile-uniform; suppose its tile type is $[x_1, x_2, \ldots, x_q]$.

By Theorem 4.1, $\mathcal{T}$ is periodic. Also, since the tiles are topological discs, we can pick a tile T and find a disc which fits inside T and a disc into which T fits. Let the diameters of these discs be u and U respectively. As all the tiles of $\mathcal{T}$ are congruent, we can fit a disc of diameter u into *any* tile, and we can fit *any* tile into a disc of diameter U. Therefore $\mathcal{T}$ obeys the little-and-large property, and so (by Theorem 2.2) we have

$$n_t(\mathcal{T}) - n_e(\mathcal{T}) + n_v(\mathcal{T}) = 0.$$

Let us rewrite this as

$$n_t(\mathcal{T}) = n_e(\mathcal{T}) - n_v(\mathcal{T}). \tag{4.1}$$

Now all the translational tile orbits of $\mathcal{T}$ consist of tiles of degree q, and therefore we can apply Theorem 3.1, with $n_q = n_t(\mathcal{T})$ and all other n_i equal to zero, to obtain

$$n_e(\mathcal{T}) = \tfrac{1}{2}(q n_t(\mathcal{T})). \tag{4.2}$$

Next, we apply Theorem 3.2, to obtain

$$n_v(\mathcal{T}) = n_t(\mathcal{T}) \left(\frac{1}{x_1} + \frac{1}{x_2} + \cdots + \frac{1}{x_q} \right). \tag{4.3}$$

Substituting Equations 4.2 and 4.3 into Equation 4.1, dividing by $n_t(\mathcal{T})$ and multiplying by 2, we obtain

$$2 = q - \left(\frac{2}{x_1} + \frac{2}{x_2} + \cdots + \frac{2}{x_q} \right).$$

As there are q bracketed terms, we can re-express this as

$$2 = \left(1 - \frac{2}{x_1}\right) + \left(1 - \frac{2}{x_2}\right) + \cdots + \left(1 - \frac{2}{x_q}\right).$$

Rather surprisingly, we now multiply by π:

$$2\pi = \left(1 - \frac{2}{x_1}\right)\pi + \left(1 - \frac{2}{x_2}\right)\pi + \cdots + \left(1 - \frac{2}{x_q}\right)\pi. \tag{4.4}$$

The significance of this is that the internal angle of a regular x_i-gon is $\left(1 - \dfrac{2}{x_i}\right)\pi$, so this equation states that the x_is must be such that we could fit the corresponding regular polygons round a point. But we saw in the proof of the Archimedean Tiling Theorem that there are just 21 ways to do this.

Theorem 2.2 of *Unit IB1*.

We have now come to the 'tedious' part of the proof — the part that we omit! In fact only 11 of the 21 possibilities for $[x_1, x_2, \ldots, x_q]$ work. These eleven types are those of the Laves tilings. ∎

Exercise 4.8

Use the solution to Exercise 3.8 to enunciate a generalization of the Archimedean Tiling Theorem.

5 THE GRÜNBAUM–SHEPHARD CLASSIFICATION

You now know that there are just eleven 'types' of transitive tiling — *if* we are content to regard two tilings as of the same type whenever they are isomorphic. Such a classification, however, ignores the rather amazing beauty and variety of transitive tilings.

At the other extreme, it would be possible to consider two transitive tilings as being 'of different types' if they differed *at all* in appearance — that is, if there were not some isometry mapping one to the other. This would clearly give us an infinity of types, parametrized by a number of continuous variables: the lengths of the edges in the different edge orbits, and so on. So, not only would a tiling isomorphic to $\mathcal{R}_4$ but by $2\,\mathrm{cm} \times 1\,\mathrm{cm}$ rectangles be (as we might expect) 'different' from $\mathcal{R}_4$, but also for every distinct (positive) x and y with (say) $x \leq y$, the tiling by $x\,\mathrm{cm} \times y\,\mathrm{cm}$ tiles would be 'different'.

This seems to be unsatisfactory. Once the *principle* that altering $\mathcal{R}_4$ into a rectangular tiling preserves its transitivity is accepted, the precise dimensions of the rectangles are of no great interest. Certain symmetries (such as rotations by $\pi/2$) are lost in going from $\mathcal{R}_4$ to a rectangular tiling, and this is the mathematically significant difference between $\mathcal{R}_4$ and any of the related rectangular tilings.

As another example, look again at tiling (b) on Side 1 of Tiling Card 4. This is isomorphic to $\mathcal{R}_6$, but more interesting! Part of the interest lies in that, despite the fact that the edges are all curved so that it certainly has not been obtained from $\mathcal{R}_6$ by an affine transformation, *some* of the symmetries of $\mathcal{R}_6$ have been preserved although some have been lost.

If you have not already sneaked a look at Tiling Cards 5–9 in the *Geometry Envelope*, you may like to do so now.

Exercise 5.1

Let $\mathcal{T}$ be tiling (b) on Side 1 of Tiling Card 4. Find:

(a) a symmetry of $\mathcal{R}_6$ which is also a symmetry of $\mathcal{T}$;

(b) a symmetry of $\mathcal{R}_6$ which is *not* a symmetry of $\mathcal{T}$.

What has effectively happened is that the straight edges of $\mathcal{R}_6$ have been replaced by circular arcs. This removes certain symmetries; but clearly the exact curvature of the arcs is not important as long as each arc has the *same* curvature.

In the next subsection, we develop these ideas into a precise mathematical definition of the *type* of a transitive tiling $\mathcal{T}$, based on the action of $\Gamma(\mathcal{T})$ on the parts of $\mathcal{T}$.

5.1 The incidence symbol of a transitive tiling

In the last exercise, you saw a tiling which is isomorphic to $\mathcal{R}_6$ but whose symmetry group is different from that of $\mathcal{R}_6$ in the sense that the two symmetry groups are not isomorphic. (The symmetry group of tiling (b) on Side 1 of Tiling Card 4 has no elements of order 6, while $\Gamma(\mathcal{R}_6)$ does have such elements.) However, the symmetry properties of tilings can differ more subtly than this: look at Figure 5.1.

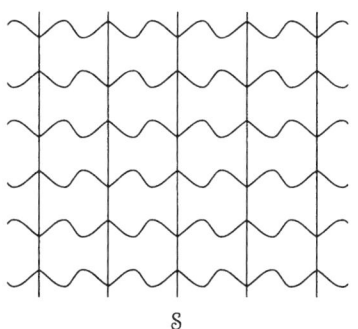 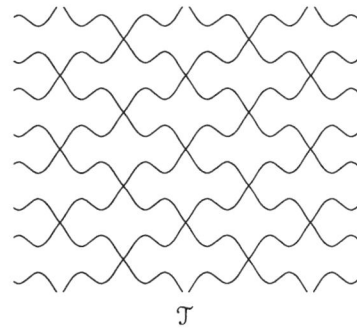

$\mathcal{S}$ $\qquad\qquad\qquad\qquad\qquad\qquad$ $\mathcal{T}$

Figure 5.1

Tilings $\mathcal{S}$ and $\mathcal{T}$ are isomorphic as each is of the tile type $[4, 4, 4, 4]$. Moreover, both are transitive, so we know from the previous section that they are both periodic; that is, $\Gamma(\mathcal{S})$ and $\Gamma(\mathcal{T})$ each contain translations in two independent directions. What about the non-translational elements of these groups? Each group has rotations through π, and each group has vertical and horizontal axes of reflection. In fact, the groups are isomorphic; but they *act* differently.

In $\mathcal{S}$, the vertical lines of reflection are along the vertical edges of the tiling, and the horizontal lines of reflection bisect the vertical edges. Thus, each vertical edge is fixed under a vertical and a horizontal reflection. As such a pair of reflections generates a group of order 4, the stabilizer of any vertical edge of $\mathcal{S}$ is of order 4. On the other hand, the stabilizer of any *tile* of $\mathcal{S}$ is generated by a horizontal reflection alone, and is thus of order 2.

Now look at $\mathcal{T}$. In this case, the lines of reflection meet at the vertices and at the centres of the tiles. No edge is fixed under any reflection, though each edge is fixed under a rotation through π about its centre. Thus the stabilizer of any edge is of order 2. Each *tile*, on the other hand, has a stabilizer of order 4.

It is also worth noting that in $\mathcal{S}$, there is one tile orbit and one vertex orbit but *three* edge orbits, while in $\mathcal{T}$, there is one tile orbit, one vertex orbit and *one* edge orbit. Thus, despite the tilings being isomorphic and having isomorphic symmetry groups, the symmetry group *actions* are quite different. We would certainly want to count these tilings as being of different types.

The method of analysis which we shall use is due to Grünbaum and Shephard, who in 1977 introduced the concept of *incidence symbols* to classify tilings and other plane patterns. The basic idea of such symbols is similar to that of tile type which you met in *Unit IB1*, in that we trace round the tile, noting how it fits in with the surrounding tiles — but in this case, we note different things.

In Section 1, you saw how to classify the edges of a tiling into edge orbits under the action of the symmetry group. Now each edge has two *sides*, corresponding to the two tiles which it separates. A side of an edge is clearly a well-defined entity, and we can consider the symmetry group to act on it.

Example 5.1

Look at Side 2 of Tiling Card 4. You will see that the shaded tile has a small arrow next to one of the edges, directed in a clockwise sense round the tile. Now imagine applying every possible element of the symmetry group; what are the possible places to which the little arrow could be shifted?

Try to answer this question for yourself before looking at Overlay 1 for Side 2 of Tiling Card 4. Then place the overlay precisely over the card, and see whether you were right. You will find a copy of the arrow against just one side of certain of the edges of the tiling. Each tile (including the marked tile) has two such arrows inside it. The arrows are labelled a, to distinguish them from other arrows that will shortly be introduced.

Now remove Overlay 1, and place Overlay 2 over the card. This places an arrow, labelled b, in the shaded tile against another of the edges. Try to think where all the other b arrows will go, then check the correctness of your guess by applying Overlay 3.

Now place Overlays 1 and 3 over the card. Some of the edges of the tiling are now labelled with an a arrow on one side and a b arrow on the other, but the work is not yet finished as some edges still lack arrows altogether. It is time for Overlay 4! Please find it and lay it over the card.

Once again, try to predict what the orbit of arrow c will look like, then verify your prediction by using Overlay 5.

This may have caught you by surprise. Each arrow is now double-headed! This is because the symmetry group contains reflections that reverse the direction of each c arrow. However, the outward curving side of each of these edges clearly cannot be mapped by a symmetry to the inward curving side (the c side), so there is still a fourth 'edge side orbit' to take into account. This is shown on Overlay 6.

Now place Overlays 1, 3, 5 and 6 all together over Side 2 of Tiling Card 4. Both sides of every edge of the tiling are now labelled and accounted for, and we have classified them into their orbits under the action of $\Gamma(\mathcal{T})$. ♦

> The two sides of an edge can be in different orbits under the action of $\Gamma(\mathcal{T})$, so we call these orbits edge *side* orbits rather than just edge orbits.

Exercise 5.2

A transitive tiling is shown in Figure 5.2. One single arrow from each edge side orbit is shown. Using a pencil (in case of error!), complete the orbits.

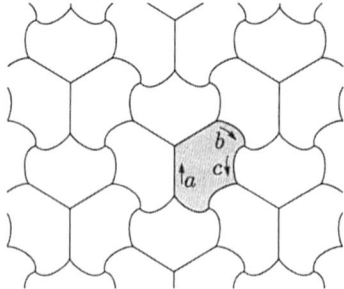

Figure 5.2

In principle, this is a reasonably complete description of the method which Grünbaum and Shephard used to find all the possible types of transitive tiling. They also extended the method to deal with vertex-transitive and edge-transitive tilings. However, there is still work to be done! A convenient *notation* is needed, so that we can refer to particular tiling types without actually drawing each tile complete with arrows.

> They use the terms *isogonal* and *isotoxal*, respectively, for these classes of tilings.

Consider tracing round a typical tile; since we are considering only transitive tilings, any tile will do. We encounter the edge side orbits in a certain order, which we can write down. In Example 5.1, for instance, these orbits are encountered in the order *abcbad* if we start at the bottom left and walk clockwise (see Figure 5.3). This, however, does not give all the information; we need to know that the first *a* and *b* are in the direction of the walk, whereas the second time we encounter them they are pointing against the direction of walk. We also need the information that *c* and *d* are double-headed arrows.

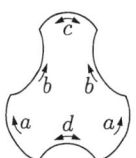

Figure 5.3

We can encode this information by using appropriately directed arrows above our letters, omitting the arrows in the double-headed case. In our case, therefore, we should write

$$\vec{a} \; \vec{b} \; c \; \cev{b} \; \cev{a} \; d.$$

This is known as the **tile symbol**.

Grünbaum and Shephard use plus and minus signs instead of arrows. Their tile symbol is therefore $a^+ \; b^+ \; c \; b^- \; a^- \; d$.

For complete information on the tiling type, we also need to know the edge orbit label on the *outside* of each edge of our typical tile. We can do this by taking a second walk around the tile, noting these outside edge labels and the directions of the arrows (see Figure 5.4). In our case, this gives

$$\vec{b} \; \vec{a} \; d \; \cev{a} \; \cev{b} \; c.$$

This is known as the **adjacency symbol**.

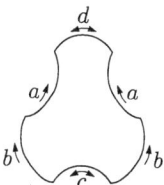

Figure 5.4

Grünbaum and Shephard construct their adjacency symbols in a slightly different way.

The tile and adjacency symbols together give the complete type information, and when expressed one after the other (or one below the other) they are known as the **incidence symbol** of the tiling. In this unit, we shall write the adjacency symbol below the tile symbol in the following way.

$$\begin{array}{cccccc} \vec{a} & \vec{b} & c & \cev{b} & \cev{a} & d \\ \vec{b} & \vec{a} & d & \cev{a} & \cev{b} & c \end{array}$$

It would be pleasant to be able to say simply that there is a one–one correspondence between incidence symbols and types of transitive tiling. However, there is the problem that the symbol generally depends on the edge from which we start our walk, and (possibly) on whether we walk clockwise or anticlockwise. In *Unit IB1* we encountered a similar problem in defining tile and vertex types. All we need do, however, is to regard two incidence symbols as meaning the same thing if one can be obtained from the other by changing the point from which the walk starts and/or the direction of walk (and also cycling or reversing the labels a, b, c, etc. as necessary). Thus, in the example we have been using, if we had started at the edge currently labelled *d* and walked anticlockwise, then the orbit currently labelled *d* would have to be relabelled *a*; the orbit currently labelled *a* would become *b*; and so on.

Exercise 5.3

What is the new incidence symbol in this case?

This would be a good stage at which to view the video programme for the unit, VC2B. The programme shows how to construct incidence symbols by thinking of a tiling in terms of a jigsaw puzzle.

5.2 Using incidence symbols

A knowledge of the topological type and the incidence symbol of a transitive tiling actually allows us to build up the complete symmetry group action. In general, this is quite a complicated operation, and you will not be required to do this; but finding the stabilizer of a tile or an edge is in fact quite straightforward, as we shall now show.

Example 5.2

Consider again the incidence symbol you met in Subsection 5.1, which is shown below.

$$\begin{array}{cccccc} \vec{a} & \vec{b} & c & \overleftarrow{b} & \overleftarrow{a} & d \\ \vec{b} & \vec{a} & d & \overleftarrow{a} & \overleftarrow{b} & c \end{array}$$

This *must* be the incidence symbol for a tiling of tile type $[3,3,3,3,3,3]$, as this is the only tile type for a transitive tiling having six edges. Can we infer anything else?

Yes, we can. The fact that the tile symbol has just two terms involving a, namely $\vec{a}$ as the first term and $\overleftarrow{a}$ as the fifth term, implies that the edge side orbit labelled a puts in just two appearances in each tile. Thus each tile is mapped to itself by the identity and just *one* other symmetry. Thus the stabilizer of a typical tile is of order 2.

We can also tell that the non-identity element of the stabilizer must be a reflection rather than a rotation: if it were a rotation, then $\vec{a}$ would map to $\vec{a}$, not $\overleftarrow{a}$. This is confirmed by looking at the other terms. The analysis for the b terms is just as for the a terms, whereas the presence of the letter c without an arrow should be interpreted as a $\vec{c}$ and a $\overleftarrow{c}$ superimposed: the c edge is mapped *to itself* by the reflection, whose axis therefore bisects the edge.

The d edge must (for similar reasons) have an axis of reflection through it. But the tile stabilizer is of order 2, so there can be only one axis of reflection through the tile, passing through the edges labelled c and d. In other words, the tile stabilizer is D_1. ♦

> Recall that the D_n and the C_n groups were described in the Appendix in *Unit IB4*. In general, D_n is the symmetry group of a regular n-gon; but for $n = 1$ or 2 this interpretation does not make sense. You should think of D_1 as the group consisting of the identity and a single reflection.

Example 5.3

Consider the transitive tiling $\mathcal{T}$ of tile type $[4,4,4,4]$, with the following incidence symbol.

$$\begin{array}{cccc} \vec{a} & \vec{b} & \vec{a} & \vec{b} \\ \overleftarrow{b} & \overleftarrow{a} & \overleftarrow{b} & \overleftarrow{a} \end{array}$$

Again, only the tile symbol (the first row of the incidence symbol) is needed for the purposes of discovering the tile stabilizer. In this case, the tile symbol contains two copies of $\vec{a}$, so again the stabilizer is of order 2; but this time there is no reversal of direction, so the stabilizer consists of the identity and a rotation through π; that is, the stabilizer is C_2. ♦

> The groups D_1 and C_2 are *isomorphic*, and considered algebraically they are both just the group $\mathbb{Z}_2$. However, they are *geometrically* distinct, as the non-identity element of D_1 is an *indirect* isometry (a reflection) whereas that of C_2 is a *direct* isometry (a rotation).

Exercise 5.4

Find the tile stabilizers of:

(a) a transitive tiling whose tile symbol is $\vec{a}\ \vec{b}\ \vec{c}\ \vec{d}$;

(b) a transitive tiling whose tile symbol is $\vec{a}\ \vec{a}\ \vec{a}\ \vec{a}$;

(c) a transitive tiling whose tile symbol is $a\ a\ a$;

(d) a transitive tiling whose tile symbol is $\vec{a}\ \overleftarrow{a}\ b\ \vec{a}\ \overleftarrow{a}\ b$.

These results exemplify the following theorem (which we shall not formally prove, although the proof is essentially contained in the foregoing discussion).

> **Theorem 5.1 Classification of tile stabilizers**
>
> Suppose we are given the tile symbol of a transitive tiling; then:
> (a) if each unarrowed letter is counted twice, the number of appearances of any given letter is the order of the tile stabilizer;
> (b) if all the letters have right arrows, the tile stabilizer is cyclic;
> (c) if not all the letters have right arrows, the tile stabilizer is dihedral.

In order to find the stabilizer of a given *edge* of a transitive tiling, all we need to do is to consider the letter in the tile symbol and the corresponding letter in the adjacency symbol. Before doing this, however, let us consider what the possibilities are. An edge E joins two vertices V and W, and any isometry that maps E to itself must *either* map V to V and W to W *or* map V to W and W to V.

Exercise 5.5

Suppose V is the point $(-1, 0)$ and W is the point $(1, 0)$.
(a) Write down the set of all isometries that map V to V and W to W.
(b) Write down the set of all isometries that map V to W and W to V.
(c) Show that the union of these sets is a group; what kind of group?

It is not difficult to see that the result of this exercise can be generalized to any two points in the plane (see Figure 5.5). Thus, the stabilizer of an edge $E = (V, W)$ must be a subgroup of the group D_2 consisting of:

the identity, e;

a reflection whose axis is the perpendicular bisector of VW; call this v;

a reflection whose axis is the straight line through V and W; call this h;

a rotation through π, about the midpoint of VW; call this r.

Therefore there are five possibilities for the stabilizer of E:
- the trivial group, $\{e\}$;
- the group $\{e, v\}$;
- the group $\{e, h\}$;
- the group $\{e, r\}$;
- the group $\{e, h, v, r\}$.

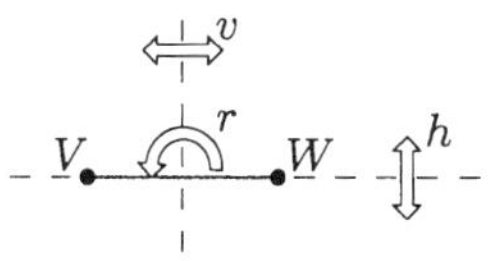

The group D_2 is just the Klein group, which you met in *Units IB2* and *IB3*, and which was there denoted by V. In the present context it is best thought of as the group generated by two reflections in lines at right angles to each other.

Figure 5.5

Now if the tile letter and the adjacency letter for an edge are the same, then the two sides of the edge can be mapped onto each other, so that the stabilizer contains either h or r or both. If they are different, then neither of these elements belongs to the stabilizer.

What about v? Clearly, this belongs to the stabilizer if the letters do not possess arrows, and does not do so if they do possess arrows.

These two observations allow us to classify the edge stabilizers completely.

> **Theorem 5.2 Classification of edge stabilizers**
>
> Suppose we are given the incidence symbol for a transitive tiling.
> (a) If the tile and adjacency letters for a given edge are different and directed (for example, $\vec{a}$ and $\vec{b}$ or $\vec{b}$ and $\overleftarrow{c}$), then the edge stabilizer is $\{e\}$.
> (b) If the tile and adjacency letters are different and undirected (for example, a and b), then the edge stabilizer is $\{e, v\}$.
> (c) If the tile and adjacency letters are the same and similarly directed (for example, both $\vec{a}$), then the edge stabilizer is $\{e, h\}$.
> (d) If the tile and adjacency letters are the same but oppositely directed (for example, $\vec{a}$ and $\overleftarrow{a}$), then the edge stabilizer is $\{e, r\}$.
> (e) If the tile letters are the same and undirected (for example, a and a), then the edge stabilizer is $\{e, h, v, r\}$.

Exercise 5.6

Classify the edge stabilizers for the tilings with the following incidence symbols.

(a) $\vec{a}\ \vec{b}\ c\ \overleftarrow{b}\ \overleftarrow{a}\ d$
$\vec{b}\ \vec{a}\ d\ \overleftarrow{a}\ \overleftarrow{b}\ c$

(b) $\vec{a}\ \vec{b}\ \vec{c}\ \vec{d}$
$\vec{c}\ \overleftarrow{b}\ \vec{a}\ \overleftarrow{d}$

(c) $\vec{a}\ \vec{a}\ \vec{a}\ \vec{a}$
$\overleftarrow{a}\ \overleftarrow{a}\ \overleftarrow{a}\ \overleftarrow{a}$

(d) $\vec{a}\ \overleftarrow{a}\ b\ \vec{a}\ \overleftarrow{a}\ b$
$\overleftarrow{a}\ \vec{a}\ b\ \overleftarrow{a}\ \vec{a}\ b$

5.3 The classification result

The fundamental result of Grünbaum and Shephard is as follows.

> **Theorem 5.3 Transitive Tiling Classification Theorem**
>
> There are exactly 81 types of transitive tiling.

B. Grünbaum and G. C. Shephard (1977) 'The eighty-one types of isohedral tilings in the plane', *Mathematical Proceedings of the Cambridge Philosophical Society*, **82**, 177–96.

As with the Archimedean Tiling Theorem in *Unit IB1*, and the Transitive Tiling Theorem in Section 4 of this unit, there is a tedious aspect to the proof of this theorem; we must write down all the *apparently* possible incidence symbols for each of the eleven types of isomorphism, and simply check which ones will in fact fit together properly. There is no need for you to go through the details of such a check. The 81 types, listed in the order produced by Grünbaum and Shephard, are shown on Tiling Cards 5–9, together with the corresponding incidence symbols. An arrow, single-headed or double-headed as appropriate, is placed in one tile against one edge in each case, to show where to start in obtaining the particular version of the incidence symbol that is given.

We have not provided overlays for Tiling Cards 5–9; instead, we have selected sixteen of the types and printed more extensive portions of these on Tiling Cards 10 and 11. These cards do have overlays to help you to explore the symmetries of the tilings.

Something you will immediately notice is that the numbers run up to 93, not 81. Nevertheless, if you count the tilings, you will find 81 of them! Twelve numbers (19, 35, 48, 60, 63, 65, 70, 75, 80, 87, 89 and 92) are missing. This is because, in the numbering scheme adopted by Grünbaum and Shephard, these numbers correspond to types of transitive tiling that can be achieved by painting motifs on the tiles to obtain certain symmetry effects that cannot be obtained by adjusting the tile shapes.

The twelve *marked tiling types* that cannot be so achieved are shown in Figure 5.6.

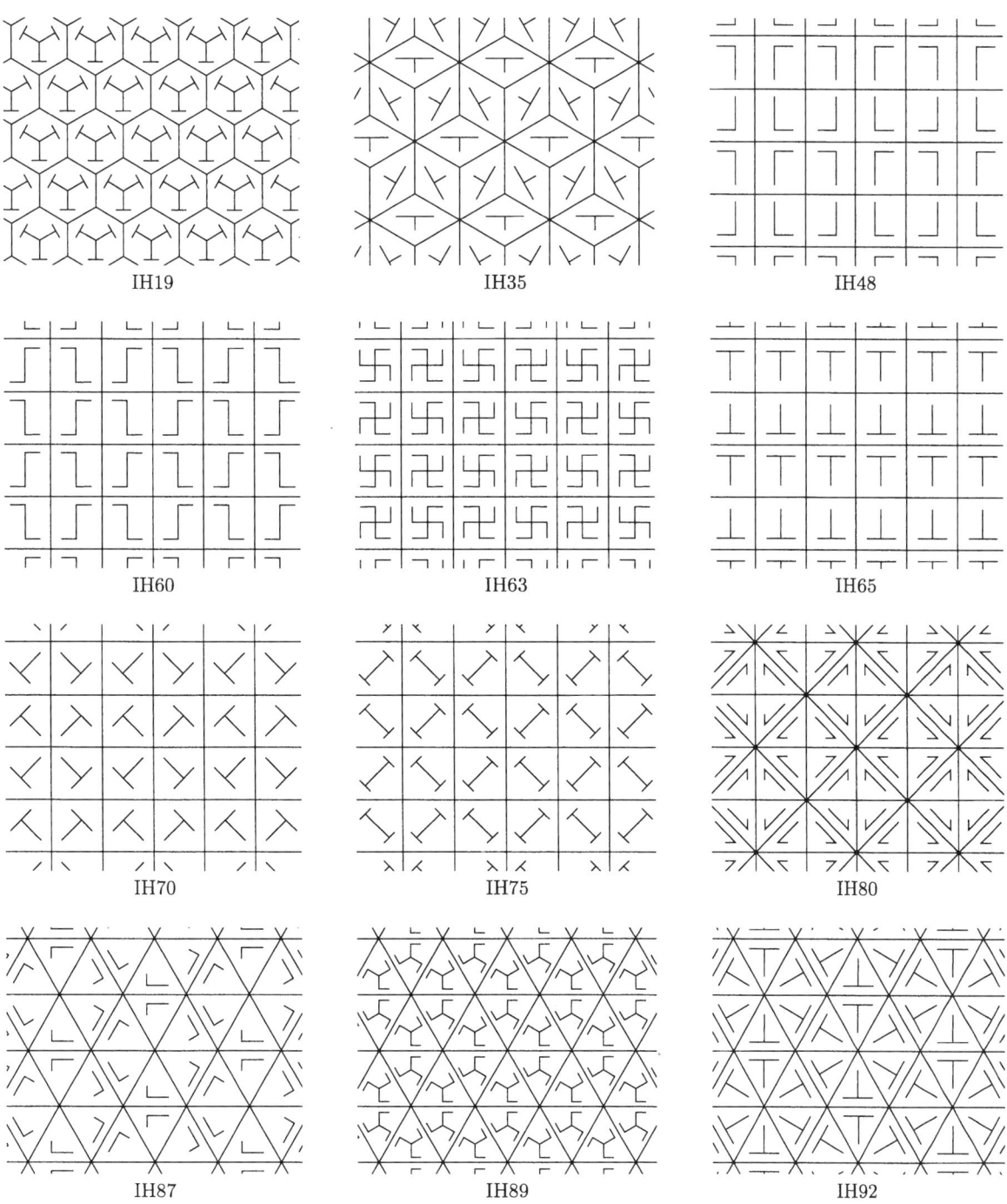

Figure 5.6

Exercise 5.7

Find the incidence symbols of the first three marked tilings above, namely IH19, IH35 and IH48.

It is important to note that two tilings (even if they are unmarked) may *look* somewhat different but still be of the same IH type in the classification scheme of Grünbaum and Shephard.

Example 5.4

Consider the tiling in Figure 5.7.

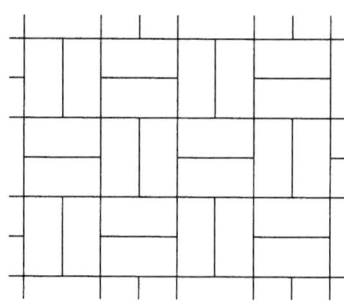

Figure 5.7

This tile type is $[3, 3, 4, 3, 4]$ and the tiling is clearly transitive, so its IH type must be 27, 28 or 29. However, if you look at the corresponding tiling card, you will *not* find a tiling all of whose edges are straight-line segments.

The point is that, in each tile, only one edge (the long edge) has $D_2 = \{e, v, h, r\}$ as stabilizer. The shorter edges, being straight, each have *individually* a symmetry group of the form D_2, but (apart from the identity) these symmetries are not symmetries *of the tiling as a whole*. Thus they are not elements of the *stabilizers* of these edges. ♦

Exercise 5.8

Using the tracing technique of Exercise 5.2 and, starting at a long edge, find the tile symbol for the tiling in Figure 5.7. Then find the adjacency symbol; hence find the IH type of the tiling.

Exercise 5.9

Classify tilings (a) and (b) in Figure 5.8.

Note that tiling (a) is by rhombuses while tiling (b) is by parallelograms that are not rhombuses — there are two different lengths of sides.

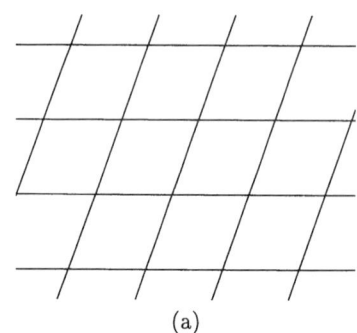 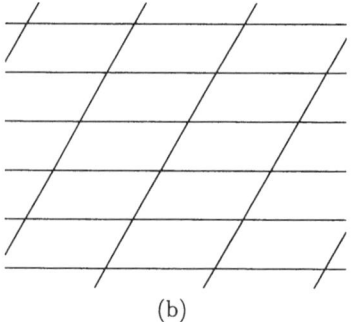

(a)   (b)

Figure 5.8

Exercise 5.10

Classify tilings (a) and (b) in Figure 5.9.

(a)　　　　(b)

Figure 5.9

These Mongolian designs are redrawn from page 11 of *Tilings and Patterns*.

APPENDIX: PROOF OF THE FINITARY THEOREM

The material in this appendix is optional.

> **Finitary Theorem**
> Any tiling which obeys the little-and-large property is finitary.

Proof

Let $\mathcal{T}$ be a tiling which obeys the little-and-large property. Let u and U be such that a disc of diameter u can be placed inside each tile of $\mathcal{T}$, and each tile of $\mathcal{T}$ can be placed inside a disc of diameter U.

Let D be any disc in the plane. We shall show that D contains or intersects only finitely many parts of $\mathcal{T}$.

Suppose D has diameter V. Consider the disc D' with the same centre as D, but with diameter $V + 2U$. Then any tile T of $\mathcal{T}$ which is contained in, intersects or touches D, lies within a disc of diameter U, which is contained in, intersects or touches D, and therefore lies entirely within D' (see Figure A.1).

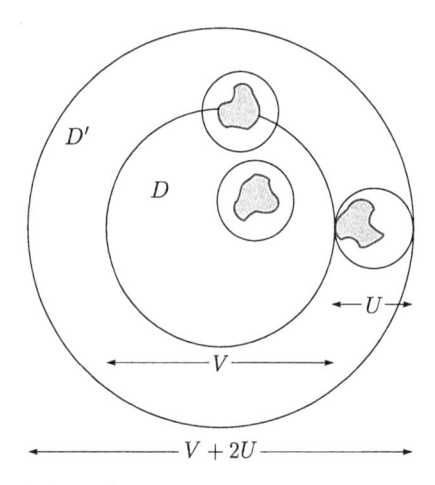

Figure A.1

Now each such tile contains a disc of diameter u, and all these discs are disjoint, and all lie within D'. Since each has area $\frac{1}{4}\pi u^2$ while V has area $\frac{1}{4}\pi(V+2U)^2$, their number cannot exceed N, where N is the largest integer not exceeding $(V+2U)^2/u^2$. Therefore D contains, intersects or touches only finitely many (at most N) tiles of $\mathcal{T}$.

A similar argument to that above shows that, for any tile T, the tiles that are adjacent to it are all contained in a disc of diameter $3U$, so there are only finitely many edges and vertices. As this is true of each one of the finitely many tiles that touch, intersect or are contained in D, it follows that D contains, intersects or touches only finitely many parts of $\mathcal{T}$ altogether. ∎

It is possible for two tiles to be adjacent along more than one edge, but with a little care it can be shown that this possibility does *not* allow a tile to be incident with *infinitely* many vertices or edges.

SOLUTIONS TO THE EXERCISES

Solution 1.1

(a): trivial;
(b), (c): generated by two independent translations;
(d): generated by one translation;
(e), (f), (g), (h): each generated by two independent translations.

Solution 1.2

The symmetry group of tiling (a) is C_6, the cyclic group of order 6 generated by $r[\pi/3]$.

Solution 1.3

All the tilings except (a) and (d) are periodic.

Solution 1.4

You need to find a way of producing progressively larger tiles as you work your way outwards. One possibility is the following.

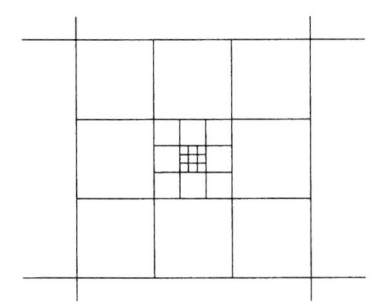

Solution 1.5

Tiling (e) is the other tiling with this property.

Solution 1.6

On Tiling Card 1, only the regular tilings $\mathcal{R}_4$ and $\mathcal{R}_6$ (that is, the Archimedean tilings $(4,4,4,4)$ and $(6,6,6)$) are translational.
On Tiling Card 3, the same two tilings are translational, and are, this time, regarded as the Laves tilings $[4,4,4,4]$ and $[3,3,3,3,3,3]$.

Solution 1.7

Tiling (b) has tile type $[3,3,3,3,3,3]$, while tiling (c) has tile type $[4,4,4,4]$.

Solution 1.8

One possibility is as follows.

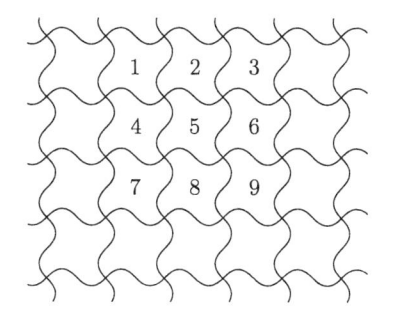

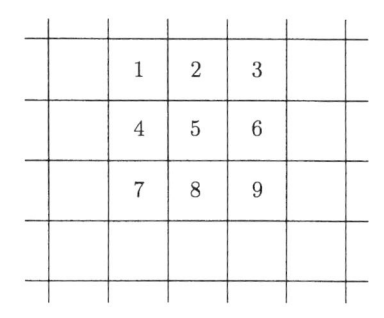

Solution 1.9

(a) Tilings (a) and (d) are isomorphic, as are tilings (b) and (c).

(b) There are several possibilities, such as the following.

For tilings (a) and (d):

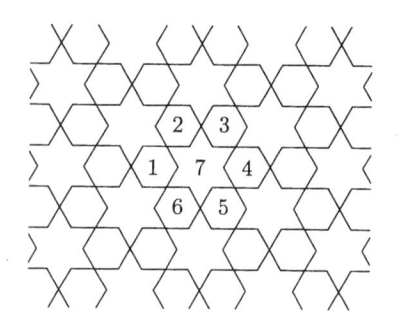

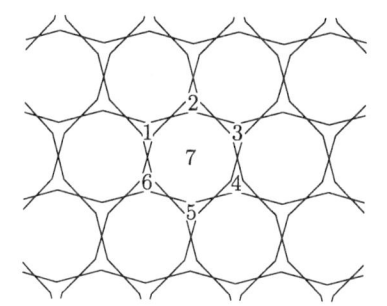

For tilings (b) and (c):

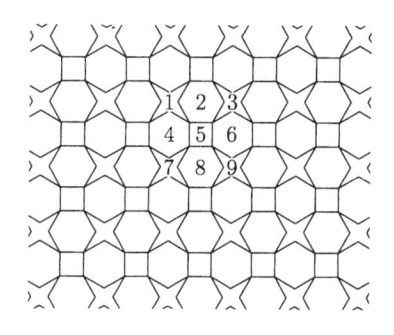

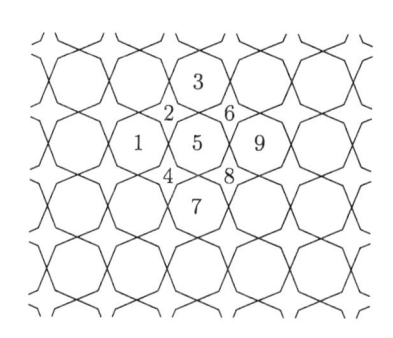

Solution 1.10

Tilings (a) and (d) of Figure 1.6 are vertex-uniform, each with vertex type $(3, 6, 3, 6)$. Therefore, by Theorem 1.2, they are isomorphic.

Tilings (b) and (c) of Figure 1.6 are vertex-uniform, each with vertex type $(4, 4, 4, 4)$. (In fact, they are also tile-uniform, with tile type $[4, 4, 4, 4]$.) Therefore, by Theorem 1.2, they are isomorphic.

Solution 1.11

Tiling Card 1

$\mathcal{T}$	$n_t(\mathcal{T})$
$(6, 6, 6)$	2
$(4, 4, 4, 4)$	1
$(3, 3, 3, 3, 3, 3)$	1
$(3, 3, 3, 3, 6)$	9
$(3, 3, 3, 4, 4)$	3
$(3, 3, 4, 3, 4)$	6
$(3, 4, 6, 4)$	6
$(3, 6, 3, 6,)$	3
$(3, 12, 12)$	3
$(4, 6, 12)$	6
$(4, 8, 8)$	2

Tiling Card 3

$\mathcal{T}$	$n_t(\mathcal{T})$
$[6, 6, 6]$	1
$[4, 4, 4, 4]$	1
$[3, 3, 3, 3, 3, 3]$	2
$[3, 3, 3, 3, 6]$	6
$[3, 3, 3, 4, 4]$	2
$[3, 3, 4, 3, 4]$	4
$[3, 4, 6, 4]$	6
$[3, 6, 3, 6]$	3
$[3, 12, 12]$	6
$[4, 6, 12]$	12
$[4, 8, 8]$	4

Solution 1.12

In each case, the equation
$$n_t(\mathcal{T}) - n_e(\mathcal{T}) + n_v(\mathcal{T}) = 0$$
holds.

Solution 2.1

Suppose an edge E is incident with vertices V and W, occupying the vector positions $\mathbf{v}$ and $\mathbf{w} \in \mathbb{R}^2$, respectively. The only translational image of E that can also be incident with W is $t[\mathbf{w} - \mathbf{v}](E)$, which translates V to W. Thus, if $n_e(\mathcal{T}) = 1$, then the degree of W must be at most 2. But we know from *Unit IB1* that any vertex of a tiling has degree at least 3. This is a contradiction. Hence $n_e(\mathcal{T}) \neq 1$.

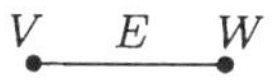

Solution 2.2

This is the case in which *both* the vertices (V, W, say) incident with E in $\mathcal{T}$ are of degree 3. If they were both in the same translational orbit, then we would have the situation depicted below.

We would have a 'tile' of infinite degree, which was infinitely long in the direction of the translational symmetry! Therefore, V and W must actually be in distinct translational orbits. When we remove E, therefore, we abolish *two* translational vertex orbits. However, the two remaining translational edge orbits incident with V coalesce into one, as do the two incident with W.

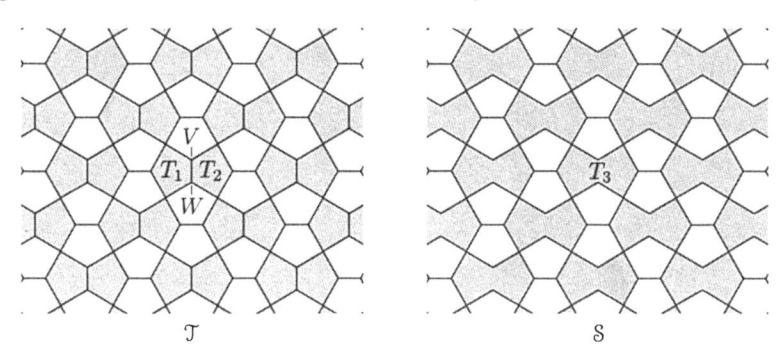

Thus,
$$n_e(\mathcal{S}) = n_e(\mathcal{T}) - 3,$$
$$n_v(\mathcal{S}) = n_v(\mathcal{T}) - 2,$$

and so we again obtain
$$n_t(\mathcal{T}) - n_e(\mathcal{T}) + n_v(\mathcal{T}) = 0.$$

Solution 3.1

The tile–edge diagram for $[3, 6, 3, 6]$ is as follows.

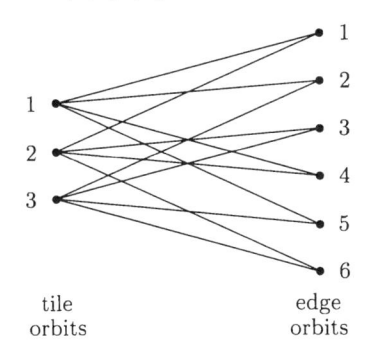

The tile–edge diagram for [4, 4, 4, 4] is as follows.

```
       •1
1 •<
       •2
tile    edge
orbit   orbits
```

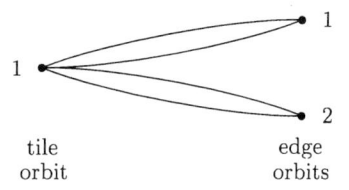

Solution 3.2

The tile–edge diagram for [3, 3, 3, 3, 3, 3] is as follows.

```
       •1
1 •<   •2
       •3
tile    edge
orbit   orbits
```

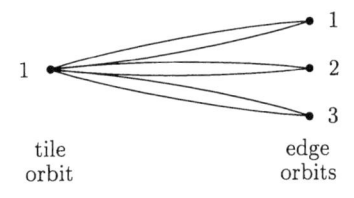

Solution 3.3

The vertex–edge diagram for [3, 6, 3, 6] is as follows.

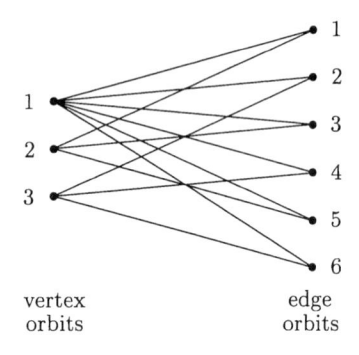

```
        •1
        •2
1 •     •3
2 •     •4
3 •     •5
        •6
vertex   edge
orbits   orbits
```

The vertex–edge diagram for [4, 4, 4, 4] is as follows.

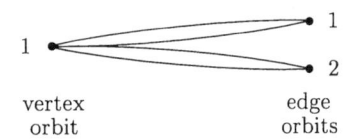

```
       •1
1 •<
       •2
vertex   edge
orbit    orbits
```

The vertex–edge diagram for [3, 3, 3, 3, 3, 3] is as follows.

```
       •1
1 •
       •2
2 •
       •3
vertex   edge
orbits   orbits
```

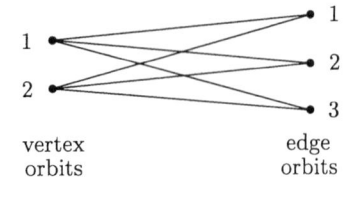

Solution 3.4

The tile–vertex diagram for $[3,3,3,4,4]$ is as follows.

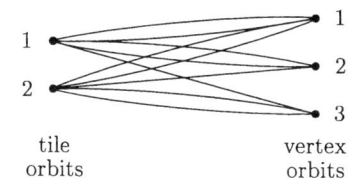

tile orbits   vertex orbits

The tile–vertex diagram for $[3,6,3,6]$ is as follows.

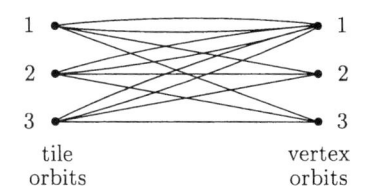

tile orbits   vertex orbits

The tile–vertex diagram for $[4,4,4,4]$ is as follows.

tile orbit   vertex orbit

Solution 3.5

The vertex–edge diagram has six dots representing vertex orbits. As each vertex is of degree 4, there are $6 \times 4 = 24$ lines going from the vertex dots to the edge dots. But each edge dot has two lines going into it, so there are $\frac{24}{2} = 12$ edge dots. Thus,

$$n_e(\mathcal{T}) = 12.$$

Solution 3.6

The tiling has been derived from $\mathcal{R}_4$ by displacing every third row. The tiles in the displaced rows form an orbit, each of whose tiles has degree 6. The tiles in the other rows form two orbits, each of whose tiles has degree 5. Thus the tile–edge diagram has $6 + (2 \times 5) = 16$ lines going from the tile dots to the edge dots, so

$$n_e(\mathcal{T}) = \tfrac{16}{2} = 8.$$

Solution 3.7

By Theorem 3.2,

$$n_v(\mathcal{T}) = 6 \left(\tfrac{1}{3} + \tfrac{1}{4} + \tfrac{1}{6} + \tfrac{1}{4} \right)$$
$$= 6.$$

Solution 3.8

Let $\mathcal{T}$ be a periodic, vertex-uniform tiling of vertex type $(x_1, x_2, \ldots, x_q)$. Then

$$n_t(\mathcal{T}) = n_v(\mathcal{T}) \left(\frac{1}{x_1} + \frac{1}{x_2} + \cdots + \frac{1}{x_q} \right).$$

Solution 4.1

The second action is transitive. (The orbits in the first action are the conjugacy classes of D_6, of which there are six.)

Solution 4.2

(a) $(4,4,4,4)$ and $(6,6,6)$.

(b) None.

(c) $(3,3,3,3,3,3)$ and $(4,4,4,4)$.

Solution 4.3

(a) Tiling Card 1: $(3,3,3,3,3,3), (4,4,4,4)$ and $(6,6,6)$;
Tiling Card 3: all;
Side 1 of Tiling Card 4: (b), (c), (f) and (g).

(b) Tiling Card 1: $(3,3,3,3,3,3), (4,4,4,4), (6,6,6)$ and $(3,6,3,6)$;
Tiling Card 3: $[3,3,3,3,3,3], [4,4,4,4], [6,6,6]$ and $[3,6,3,6]$;
Side 1 of Tiling Card 4: (b), (c) and (g).

(c) Tiling Card 1: all;
Tiling Card 3: $[3,3,3,3,3,3], [4,4,4,4]$ and $[6,6,6]$;
Side 1 of Tiling Card 4: (b), (c), (f) and (g).

Solution 4.4

In terms of the notation of the Isometry Toolkit, we have

$$\theta_P = r[\pi/2];$$
$$\begin{aligned}\theta_Q &= r[(1,0), \pi/2] \\ &= t[(1,0) - (0,1)]\, r[\pi/2] \quad \text{(by Equation 8 of the Toolkit)} \\ &= t[(1,-1)]\, r[\pi/2];\end{aligned}$$

similarly,

$$\begin{aligned}\theta_R &= r[(1,1), \pi/2] \\ &= t[(1,1) - (-1,1)]\, r[\pi/2] \\ &= t[(2,0)]\, r[\pi/2].\end{aligned}$$

Therefore,

$$\begin{aligned}\theta_Q^{-1}\theta_P &= r[-\pi/2]\, t[(-1,1)]\, r[\pi/2] \\ &= t[(1,1)]\, r[-\pi/2]\, r[\pi/2] \quad \text{(by Equation 6 of the Toolkit)} \\ &= t[(1,1)];\end{aligned}$$

$$\begin{aligned}\theta_R^{-1}\theta_P &= r[-\pi/2]\, t[(-2,0)]\, r[\pi/2] \\ &= t[(0,2)]\, r[-\pi/2]\, r[\pi/2] \\ &= t[(0,2)].\end{aligned}$$

Solution 4.5

(a) Using the Isometry Toolkit as in Exercise 4.4, we obtain

$$\begin{aligned}\theta_P &= r[\mathbf{p}, \theta] \\ &= t[\mathbf{p}']\, r[\theta],\end{aligned}$$

where

$$\mathbf{p}' = \mathbf{p} - r[\theta](\mathbf{p}).$$

Similarly,

$$\theta_Q = t[\mathbf{q}']\, r[\theta] \quad \text{and} \quad \theta_R = t[\mathbf{r}']\, r[\theta],$$

where

$$\mathbf{q}' = \mathbf{q} - r[\theta](\mathbf{q}) \quad \text{and} \quad \mathbf{r}' = \mathbf{r} - r[\theta](\mathbf{r}).$$

Thus,

$$\begin{aligned}\theta_Q^{-1}\theta_P &= r[-\theta]\, t[-\mathbf{q}']\, t[\mathbf{p}']\, r[\theta] \\ &= r[-\theta]\, t[\mathbf{p}' - \mathbf{q}']\, r[\theta].\end{aligned}$$

Using Equation 6 of the Toolkit as in Exercise 4.4, this is a translation by $r[-\theta](\mathbf{p}' - \mathbf{q}')$. Now,
$$r[-\theta](\mathbf{p}') = r[-\theta](\mathbf{p} - r[\theta](\mathbf{p}))$$
$$= r[-\theta](\mathbf{p}) - \mathbf{p}.$$

Similarly, $r[-\theta](\mathbf{q}') = r[-\theta](\mathbf{q}) - \mathbf{q}$, and so finally
$$\theta_Q^{-1}\theta_P = t[\mathbf{a}],$$
where
$$\mathbf{a} = r[-\theta](\mathbf{p}) - \mathbf{p} - r[-\theta](\mathbf{q}) + \mathbf{q}$$
$$= \mathbf{q} - \mathbf{p} - r[-\theta](\mathbf{q} - \mathbf{p})$$
$$= \mathbf{A}(\mathbf{q} - \mathbf{p}),$$
where
$$\mathbf{A} = \begin{bmatrix} 1 - \cos(-\theta) & \sin(-\theta) \\ -\sin(-\theta) & 1 - \cos(-\theta) \end{bmatrix}$$
$$= \begin{bmatrix} 1 - \cos\theta & -\sin\theta \\ \sin\theta & 1 - \cos\theta \end{bmatrix}.$$

By a similar argument,
$$\theta_R^{-1}\theta_P = t[\mathbf{b}],$$
where
$$\mathbf{b} = \mathbf{A}(\mathbf{r} - \mathbf{p}).$$

(b) $\det \mathbf{A} = (1 - \cos\theta)^2 + \sin^2\theta$
$$= 1 - 2\cos\theta + \cos^2\theta + \sin^2\theta$$
$$= 2(1 - \cos\theta)$$
$$\neq 0 \quad \text{if } \theta \text{ is not a multiple of } 2\pi.$$

Therefore, since $\mathbf{q} - \mathbf{p}$ and $\mathbf{r} - \mathbf{p}$ are linearly independent, so are $\mathbf{A}(\mathbf{q} - \mathbf{p}) = \mathbf{a}$ and $\mathbf{A}(\mathbf{r} - \mathbf{p}) = \mathbf{b}$.

Solution 4.6

Using Equation 15 of the Toolkit, we obtain
$$g_1 = q[(1,0),(0,1),0] = t[(1,0) + 2(0,1)]\, q[0] = t[(1,2)]\, q[0],$$
$$g_2 = q[(1,0),(0,2),0] = t[(1,0) + 2(0,2)]\, q[0] = t[(1,4)]\, q[0],$$

and so
$$g_1^2 = t[(1,2)]\, q[0]\, t[(1,2)]\, q[0]$$
$$= t[(1,2)]\, t[(1,-2)]\, (q[0])^2$$
$$= t[(2,0)],$$

$$g_2 g_1 = t[(1,4)]\, q[0]\, t[(1,2)]\, q[0]$$
$$= t[(1,4)]\, t[(1,-2)]\, (q[0])^2$$
$$= t[(2,2)].$$

Solution 4.7

Using Equation 15 of the Toolkit, we obtain

$$g_1 = t[\mathbf{g} + 2\mathbf{c}]q[\theta], \quad g_2 = t[\lambda\mathbf{g} + 2\mu\mathbf{c}]\, q[\theta],$$

and so

$$\begin{aligned}g_1^2 &= t[\mathbf{g} + 2\mathbf{c}]\, q[\theta]\, t[\mathbf{g} + 2\mathbf{c}]\, q[\theta] \\ &= t[\mathbf{g} + 2\mathbf{c}]\, t[\mathbf{g} - 2\mathbf{c}]\, (q[\theta])^2 \\ &= t[2\mathbf{g}],\end{aligned}$$

$$\begin{aligned}g_2 g_1 &= t[\lambda\mathbf{g} + 2\mu\mathbf{c}]\, q[\theta]\, t[\mathbf{g} + 2\mathbf{c}]\, q[\theta] \\ &= t[\lambda\mathbf{g} + 2\mu\mathbf{c}]\, t[\mathbf{g} - 2\mathbf{c}]\, (q[\theta])^2 \\ &= t[(1 + \lambda)\mathbf{g} + 2(\mu - 1)\mathbf{c}].\end{aligned}$$

Since $\mathbf{g} \neq \mathbf{0}$, the translation $t[2\mathbf{g}]$ is a non-zero translation, and since $\mu \neq 1$ and $\mathbf{c} \neq \mathbf{0}$, the translation $t[(1 + \lambda)\mathbf{g} + 2(\mu - 1)\mathbf{c}]$ is also a non-zero translation and is in a linearly independent direction from $2\mathbf{g}$.

Solution 4.8

Solution 3.8 shows that all the work of Subsection 4.3 can be done with vertex and edge orbits interchanged. Therefore, we have a theorem analogous to Theorem 4.2:

> Every vertex-transitive tiling is isomorphic to one of the eleven Archimedean tilings.

Solution 5.1

(a) All the translational symmetries of $\mathcal{R}_6$ (for example) are also symmetries of $\mathcal{T}$, as are the rotations through $\pm 2\pi/3$ about the vertices, and about the centres of the tiles.

(b) Any rotation through $\pm\pi/3$ or through π about the centre of a tile of $\mathcal{R}_6$ does not correspond to any symmetry of $\mathcal{T}$.

Solution 5.2

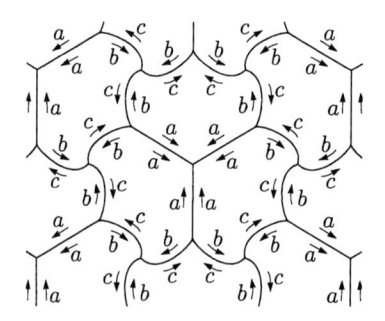

Solution 5.3

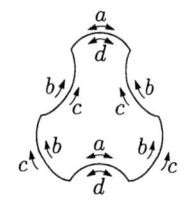

$$\begin{array}{cccccc} a & \vec{b} & \vec{c} & d & \overleftarrow{c} & \overleftarrow{b} \\ d & \vec{c} & \vec{b} & a & \overleftarrow{b} & \overleftarrow{c} \end{array}$$

Solution 5.4

(a) The tile symbol contains four *different* symbols, so the tile has no symmetry. Thus the tile stabilizer is $\{e\}$, which we regard as the cyclic group C_1.

We regard the trivial group $\{e\}$ as the cyclic group C_1.

(b) The tile symbol has four terms involving $\vec{a}$, so the tile stabilizer is a rotation group of order 4, namely C_4.

(c) The tile symbol has three terms involving a, so the tile stabilizer contains 3 rotations and 3 reflections and is therefore D_3.

(d) The tile symbol contains $\vec{a}$ twice and $\overleftarrow{a}$ twice, so the tile stabilizer is a group of order 4 containing 2 rotations and 2 reflections, and is therefore D_2.

Solution 5.5

(a) The identity, and reflection in the x-axis.
(b) Reflection in the y-axis, and rotation through π about the origin.
(c) The composite of the two reflections is the rotation; the composite of either reflection with the rotation is the other reflection. The group is D_2.

Solution 5.6

(a) The first, second, fourth and fifth edges are in the same edge orbit and have stabilizer $\{e\}$; the third and sixth edges are in the same orbit and have stabilizer $\{e, v\}$.

(b) The first and third edges are in the same edge orbit, and have edge stabilizer $\{e\}$; the second and fourth edges are in separate orbits and each has stabilizer $\{e, r\}$.

(c) These edges are all in the same edge orbit and have stabilizer $\{e, r\}$.

(d) The first, second, fourth and fifth edges are in the same edge orbit and have stabilizer $\{e, r\}$; the third and sixth edges are in the same edge orbit and have stabilizer $\{e, h, v, r\}$.

Solution 5.7

The incidence symbol for IH19 is as follows.

$$\begin{array}{cccccc} \vec{a} & \overleftarrow{a} & \vec{a} & \overleftarrow{a} & \vec{a} & \overleftarrow{a} \\ \vec{a} & \overleftarrow{a} & \vec{a} & \overleftarrow{a} & \vec{a} & \overleftarrow{a} \end{array}$$

The incidence symbol for IH35 is as follows.

$$\begin{array}{cccc} \vec{a} & \vec{b} & \overleftarrow{b} & \overleftarrow{a} \\ \vec{a} & \vec{b} & \overleftarrow{b} & \overleftarrow{a} \end{array}$$

The incidence symbol for IH48 is as follows.

$$\begin{array}{cccc} \vec{a} & \vec{b} & \vec{c} & \vec{d} \\ \vec{a} & \vec{b} & \vec{c} & \vec{d} \end{array}$$

Solution 5.8

The tracing process gives the following orbits.

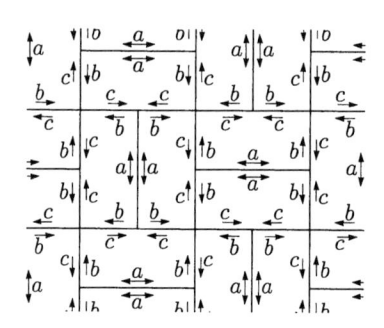

[figure]

Thus, the tile symbol is

$$\vec{a} \quad \vec{b} \quad \overset{\leftarrow}{c} \quad \overset{\leftarrow}{c} \quad \overset{}{b}$$

and the adjacency symbol is

$$\overset{}{a} \quad \overset{\leftarrow}{c} \quad \overset{\leftarrow}{b} \quad \vec{b} \quad \vec{c}$$

and so the tiling is of type IH29.

Solution 5.9

Tiling (a) has the incidence symbol

$$\vec{a} \quad \overset{\leftarrow}{a} \quad \vec{a} \quad \overset{\leftarrow}{a}$$
$$\overset{\leftarrow}{a} \quad \vec{a} \quad \overset{\leftarrow}{a} \quad \vec{a}$$

and so the tiling is of type IH76.

Tiling (b) has the incidence symbol

$$\vec{a} \quad \vec{b} \quad \vec{a} \quad \vec{b}$$
$$\overset{\leftarrow}{a} \quad \overset{\leftarrow}{b} \quad \overset{\leftarrow}{a} \quad \overset{\leftarrow}{b}$$

and so the tiling is of type IH57.

Solution 5.10

(a) The tile type is $[3,3,3,3,3,3]$, and the incidence symbol is as follows.

$$a \quad b \quad a \quad b \quad a \quad b$$
$$b \quad a \quad b \quad a \quad b \quad a$$

So the tiling is of type IH18.

(b) The tile type is again $[3,3,3,3,3,3]$, and the incidence symbol is as follows.

$$a \quad \vec{b} \quad \overset{\leftarrow}{b} \quad a \quad \vec{b} \quad \overset{\leftarrow}{b}$$
$$a \quad \overset{\leftarrow}{b} \quad \vec{b} \quad a \quad \overset{\leftarrow}{b} \quad \vec{b}$$

So the tiling is of type IH17.

OBJECTIVES

After working through this unit, you should be able to:

(a) explain what is meant by a periodic tiling and by a translational tiling;

(b) understand the little-and-large property as applied to tilings;

(c) find the tile, edge and vertex orbits under the action of the translation group of a tiling;

(d) understand the proof of the fact that every translational tiling is tile-uniform, of tile type $[4,4,4,4]$ or $[3,3,3,3,3,3]$;

(e) know and understand the proof of the Euler Equation for periodic tilings obeying the little-and-large property;

(f) construct the tile–edge, the vertex–edge and the tile–vertex diagrams of a given periodic tiling;

(g) use the properties of these diagrams to establish the Edge Orbit and Vertex Orbit Theorems for periodic tilings;

(h) explain what is meant by a transitive tiling;

(i) prove that every transitive tiling is periodic, and understand the proof that every transitive tiling is isomorphic to a Laves tiling;

(j) find the tile and edge stabilizers of a given transitive tiling;

(k) understand the outlines of the method devised by Grünbaum and Shephard for classifying transitive tilings into 81 types.

ACKNOWLEDGEMENT

Grateful acknowledgement is made to the following source for permission to reproduce material in this unit:

Page 35: Grünbaum, B. and Shephard, G. C. (1977) 'The eighty-one types of isohedral tilings in the plane', *Mathematical Proceedings of the Cambridge Philosophical Society*, **82**, Cambridge University Press.

INDEX

adjacency symbol 31
classification of edge stabilizers 34
classification of tile stabilizers 33
edge orbit theorem 21
Euler equation 14
finitary 8
finitary theorem 8
frieze group 7
incidence symbol 31
isomorphic (tilings) 10
isomorphism (of tilings) 10

little-and-large property 8
marked tiling types 35
orbit diagrams 18
periodic tiling 7
point group 7
tile–edge diagram 18
tile–vertex diagram 20
tile-transitive 24
tile symbol 31
transitive group action 23
transitive tiling 24

transitive tiling classification
  theorem 34
transitive tiling theorem 27
translational edge orbits 12
translational tile orbits 12
translational tiling 9
translational vertex orbits 12
translation group 6
vertex–edge diagram 19
vertex orbit theorem 22
wallpaper group 7